STUDENT UNIT GUIDE

OCR AS A2

UNITS
F213
F216

Biology

Practical Skills in Biology

Richard Fosbery

ALLAN

The author wishes to thank Jenny Wakefield-Warren, Jacky Williams and Keith Hirst for their comments on the manuscript and Stephanie Fowler for ideas for the Skills Guidance section.

First published in 2010 by Philip Allan Updates, an imprint of Hodder Education, an Hachette UK company, Market Place, Deddington, Oxfordshire OX15 0SE

Orders

Bookpoint Ltd, 130 Milton Park, Abingdon, Oxfordshire OX14 4SB
tel: 01235 827827
fax: 01235 400401
e-mail: education@bookpoint.co.uk

Lines are open 9.00 a.m.–5.00 p.m., Monday to Saturday, with a 24-hour message answering service. You can also order through the Philip Allan Updates website: www.philipallan.co.uk

© Richard Fosbery 2010
ISBN 978-1-4441-1546-8

First printed 2010
Impression number 5 4 3 2 1
Year 2014 2013 2012 2011 2010

This guide has been written specifically to support students preparing for the OCR Biology Units F213 (AS) and F216 (A2). The content has neither been approved nor endorsed by OCR and remains the sole responsibility of the author. Information about practical tasks and the student answers provided are for illustrative purposes only.

Printed by MPG Books, Bodmin

Hachette UK's policy is to use papers that are natural, renewable and recyclable products and made from wood grown in sustainable forests. The logging and manufacturing processes are expected to conform to the environmental regulations of the country of origin.

Contents

Introduction

■ ■ ■

Skills Guidance

F213: Practical work at AS

F216: Practical work at A2

■ ■ ■

Practical tasks

Introduction

About this guide

This unit guide is intended to help you prepare for **Unit F213: Practical Skills in Biology 1** and **Unit F216: Practical Skills in Biology 2**. It is divided into three sections:

- **Introduction** — this gives advice on how to use the guide to help your preparation for the skills assessment and the three tasks in each unit.
- **Skills Guidance** — this is a guide to the skills, procedures and terminology involved in the practical work on which you will be assessed. The terminology of practical work in biology is defined and explained in detail. There is also detailed advice on the **qualitative task**, the **quantitative task** and the **evaluative task**. All the practical skills that you need are discussed in the context of some practicals that you are likely to do at AS. In the second half there is a discussion of the practical work you are likely to do at A2, including fieldwork for Module 3 of Unit F215. The differences between AS and A2 tasks are explained. The preparation you do for the practical assessment is also useful for the other unit tests.
- **Practical Tasks** — here you will find three tasks for Unit F213 and three tasks for Unit F216, together with answers written by two candidates and examiner's comments. There is also more advice about how to prepare for the practical tasks.

This guide is not a substitute for carrying out practical work in class and should be read alongside the work that you do in the laboratory. Practical work should be informative and enjoyable. However, if you have difficulties with any of the laboratory skills required, it can be frustrating. If this is the case then you should ask your teachers for help. The biology technicians may be able to help with the use of specific equipment or particular procedures. Throughout the guide there are numerical problems and questions that you should try. These are indicated by 'Try this yourself' with the answers on pages 123–27. These questions will give you some idea of the types of question the examiners may ask, particularly on the theory behind the tasks. The work you do in preparing for Units F213 and F216 will also help you with the examinations for the other units. There are learning outcomes on practical work and details of laboratory procedure, so data analysis, interpretation and evaluation are assessed in the unit tests, as well as in the practical tasks. Make sure that you write notes on the practical work to accompany your notes on the rest of the course.

The **Skills Guidance** section will help you to:

- understand what you are expected to do in the three tasks in each unit
- check that you understand the links to theory, since you need these to interpret the data you collect

- understand what is meant by each of the main skills
- respond correctly to the terminology used by examiners
- use correct terminology in your answers and avoid confusing the key terms

The main aspects of the practical tasks are explained by the use of examples drawn from the specification. These cover many of the topics likely to be used as contexts for the practical tasks.

The **Practical Tasks** section will help you to:

- check the way examiners prepare tasks and ask questions
- understand what the examiners mean by terms such as 'reliable', 'accurate', 'precise', 'confident' and 'valid'
- interpret the question material, especially any data that the examiners give you
- write concisely and answer the questions that the examiners set

Practical assessment in AS and A2 biology

The specification outlines what you are expected to learn and do. The content of the specification is written as **learning outcomes**; these state what you should be able to do after studying and revising each topic. Some learning outcomes refer to practical work and it is these that are likely to form the basis of the practical assessment. They could also be examined in the theory papers. Make sure you understand fully what you are required to do in response to these learning outcomes. The specification also lists some ideas for practical work at the end of each module. Unit F213 contributes 20% to the AS grade and 10% to the full A-level grade. Unit F216 contributes 10% to the A-level grade.

Your school or college will decide when you will carry out the practical tasks. They are done in class time. You may not take any of the work away from class; you may not do any of it in study time or at home. Examiners write the tasks and compile the mark schemes. Your teachers will mark your answers and retain your work in case it has to be submitted to OCR as part of a sample of their marking. Moderators appointed by OCR check the marking and decide whether the marks awarded by your teacher could possibly be increased or decreased. Do not assume that the mark you have been given is the one that OCR will use to aggregate with the marks from your other unit tests to give your final grade.

Examiners prepare new tasks each year. It is usual to start the tasks in the first term of AS, although some schools and colleges may leave them until after Christmas. Some centres do tasks on ecology for F216 (A2) on field trips in the summer term after the AS examinations. You should find out how many opportunities you will have to take the tasks. Your teacher will submit a single mark for each type of task to give a mark out of a total of 40. He/she will pick the best mark for each type of task if you have completed more than one. Your marks for Units F213 and F216 will be submitted in May of your AS and A2 years. If you wish to retake these units to improve your marks, you must submit at least one task not previously submitted. Your marks for the other tasks may be resubmitted without further work on your part, but your teacher must remark the work.

The practical tasks

Each task is printed in a booklet, in which you write all your answers. There is no time limit for the qualitative and quantitative tasks but they have to be completed in class time.

The qualitative task is completed in one practical session, which is likely to be about 1 hour.

The quantitative task is divided into two parts. In Part 1, you carry out the procedure in the laboratory to collect some quantitative data. In Part 2, you process, analyse and interpret the data that you collected in Part 1. Your teachers are not permitted to give you any data if you have problems with your practical work or if you think your results are 'wrong'. You will still be awarded marks in Part 2 for processing, presenting and analysing your data, so you must complete Part 2 with whatever data you have collected. You should be able to complete each part in about 1 hour.

The evaluative task is based on the quantitative task, but using data provided by the examiners. The questions ask you to do further processing, suggest limitations of the procedure and ways in which it could be improved. The evaluative task may ask you how the limitations affect the data and ask you to explain your improvements to the procedure. You will also be asked to evaluate the quality of the data, comment on key definitions and possibly use or comment on more complex processing such as statistics. This task must be completed in 1 hour unless you are allowed extra time in examinations. You will also have to use your knowledge and understanding of the topics to make conclusions. Expect to do this in both the quantitative and evaluative tasks.

In the quantitative and evaluative tasks you are expected to use numeracy skills and perform calculations. You should have a calculator when you carry out these tasks; make sure you know how to perform the various calculations expected of you. As the tasks are carried out in the laboratory under timed conditions you may not have access to a computer to present data or perform these calculations.

All your answers are written on the question papers. You should have a pen, pencil, ruler, eraser, pencil sharpener and calculator when you take each of the tests. You are not allowed to have any books, notes or help sheets.

There are 40 marks available for F213 and F216. The marks are awarded as shown in the table.

Task	Teacher observation	Teacher marking of written work	Total marks
Qualitative task	Safe working (1 mark) Skilful practice (1 mark)	Recording results Making observations Answering questions (8 marks)	10
Quantitative task	Skilful practice (2 marks)	Answering questions (8 marks)	10
Evaluative task		Presenting processed data Analysing and interpreting processed data Evaluating procedures and data	20

Units of measurement

It is important that you know the units of measurement used in practical work. You will also calculate derived units such as for rates. There is advice about the correct units to use and how to carry out calculations throughout this guide. The table below gives you the units you are most likely to use in your practical work.

Unit	Name of unit	Symbol	Notes
Length	Metre Centimetre Millimetre Micrometre	m cm (10^{-2} m) mm (10^{-3} m) μm (10^{-6} m)	The centimetre is not an SI unit, so most measurements are given in millimetres; you may be expected to calculate actual sizes of cells using the micrometre
Volume	Decimetre cubed Centimetre cubed	dm^3 cm^3	These are the units used in examination papers, although you may well find litres (l or L) or millilitres (ml) on apparatus such as glassware and syringes
Mass	Kilogram Gram	kg g (10^{-3} kg)	
Amount of substance	Mole Millimole Micromole	mol mmol (10^{-3} mol) μmol (10^{-6} mol)	Concentrations may be given as $mol\,dm^{-3}$ (often said as mol per litre) or as mass per volume, e.g. $g\,dm^{-3}$ (grams per dm^3); you will also see percentage solutions, which are $g\,100\,cm^{-3}$ (grams per 100 cm^3)
Temperature	Degrees Celsius	°C	
Pressure	Pascal Kilopascal Megapascal	Pa kPa (10^3 Pa) MPa (10^3 kPa)	Pressure units are used for water potential, solute potential and pressure potential in studies on osmosis

Note: SI stands for *Système Internationale d'Unités*. There are seven **base units**: the metre, the kilogram, the second, the ampere, the kelvin, the mole, and the candela. **Derived units** are those formed by combining base units according to the algebraic relations linking the corresponding quantities (for more information on this see **www.bipm.org/en/si**).

Common examples of derived units are those used for rates of reaction. For example, you may measure the appearance of a product during an enzyme-catalysed reaction. If this is a gaseous product, you will measure its volume in cm^3. To calculate the rate of reaction you divide the volume collected by the time taken, in minutes or seconds, to collect it. This gives you a rate in cm^3 per minute or cm^3 per second. These derived units are written using negative exponents:
- $cm^3 min^{-1}$
- $cm^3 s^{-1}$

There is more about this on page 20.

Names of organisms

In Unit F212 you learnt how to name organisms using the binomial system. The scientific names of organisms are used throughout this guide, with common names where appropriate. This is important in sections on biodiversity, ecology and fieldwork at A2, where you may have to refer to the organisms that you have collected, sampled and identified.

Skills guidance

This section is a guide to the skills that you need for Unit F213: Practical Skills in Biology 1 and Unit F216: Practical Skills in Biology 2.

In this guide you will find references to the three tasks that you will carry out in Unit F213 and Unit F216. They are:

- the **qualitative task**, e.g. implementing a practical procedure that does not give you anything you can measure or determine — it may involve recording colours, counting specimens, making identifications or drawing from a specimen or from a microscope slide.
- the **quantitative task**, e.g. implementing a procedure in which you record measurements; it may also involve the processing and presentation of results
- the **evaluative task**, e.g. commenting critically on the practical procedure and the results you obtained while doing the quantitative task. You will also be given some more results to process and comment on. You do not carry out any more practical work for this task

The skills assessed are summarised below.

Qualitative task

(a) demonstrate skilful and safe practical techniques using suitable **qualitative** methods

(b) make and record valid observations

Quantitative task

(a) demonstrate skilful and safe practical techniques using suitable **quantitative** methods

(b) make and record accurate measurements to an appropriate degree of precision

(c) analyse, interpret and evaluate experimentally derived results quantitatively to reach valid conclusions

Evaluative task

(a) process results quantitatively

(b) analyse and interpret data, identify anomalies and reach valid conclusions

(c) assess the reliability and accuracy of an experimental task

(d) identify and explain the main limitations of the data collection strategy and identify weaknesses in experimental procedures and measurements

(e) understand and propose simple improvements to experimental procedures and measurements

The examples used throughout this section are based on topics from the learning outcomes or the practical suggestions given at the end of each module in the specification. There are full lists of these in Table 1 for AS (page 11) and Table 23 for A2 (pages 55–56).

F213: Practical work at AS

The practical work that you do throughout your AS course will help to develop the skills that are assessed in the tasks. This part of the guide uses examples of practicals that are relevant to the content of the two AS units: F211 and F212. Table 1 is a checklist of the examples used for Unit F213.

Table 1 Topics from AS Units F211 and F212 used for examples of practical tasks

Example		Unit	Module(s)	Task	Skill
1	Leaf shape and keys	F212	2.3.2	Qualitative	(b)
2	Leaf shape, types of variable and types of data	F212	2.3.2	Qualitative	(b)
3	Red blood cells and osmosis	F211	2.1.2	Qualitative	(a) (b)
4	Sensitivity of Benedict's test	F212	2.1.1	Qualitative	(a) (b)
5	Inhibition of amylase	F212	2.1.3	Qualitative	(a) (b)
6	Observing plasmolysis	F211	1.1.2 and 1.2.3	Qualitative	(b)
7	Drawing a tissue map	F211	1.1.3 and 1.2.3	Qualitative	(b)
8	Drawing plant cells	F211	1.1.2 and 1.2.3	Qualitative	(b)
9	Loss of chloride ions from plant tissue	F211	1.1.2	Semi-quantitative	(b)
10	Colour standards with Benedict's solution	F212	2.1.1	Semi-quantitative	(a) (b)
11	Decolourising potassium manganate(VII)	F212	2.1.1	Quantitative	(a) (b)
12	Rate of diffusion in agar blocks	F211	1.2 and 1.2.1	Quantitative Evaluative	(b) (c) (b)
13	Mass change of potato in sucrose solutions	F211	1.1.2 and 1.2.3	Quantitative Evaluative	(b) (c) (a) (b) (c)
14	Hydrolysis of starch	F212	2.1.3	Quantitative Evaluative	(b) (c) (a)
15	Decomposition of hydrogen peroxide	F212	2.1.3	Quantitative Evaluative	(b) (c) (a)
16	Effect of pH on amylase	F212	2.1.3	Quantitative Evaluative	(b) (c) (a)
17	Catalase in various plant tissues	F212	2.1.3	Quantitative Evaluative	(b) (c) (a)
18	Catalase and hydrogen peroxide	F212	2.1.3	Quantitative Evaluative	(b) (c) (d)

Investigations in biology usually begin with observations of the natural world. These observations prompt questions that in turn prompt hypotheses. Scientists design experiments to test hypotheses. Qualitative tasks test your ability to make observations, record and interpret them. Quantitative tasks involve collecting numerical data that you process by carrying out calculations and present in the form of charts or graphs.

Qualitative tasks: general

Example 1: leaf shape and keys

This example is about making observations.

Figure 1 shows outline drawings of ten leaves. Look carefully at the leaves and make a list of the different features that you can see. Include in your list all the features you can see even if some are shown by only one of the leaves. The features you are listing are **qualitative** features. If you could measure the lengths and widths of each leaf and find out its mass then you could include some **quantitative** data.

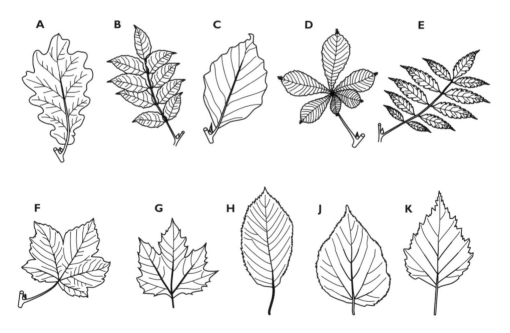

Figure 1 Outline drawings of leaves from ten species of tree

The data that you have collected about the leaves have to be organised. The best way to organise information is in a table. Some of the features that you might have identified are in given in Table 2.

Table 2 Partly completed table of observations about the leaves shown in Figure 1

Leaf	Features				
	Compound leaf	Simple leaf	Toothed margin	Midrib with one main vein	Petiole (leaf stalk)
A	✗	✓	✗	✓	✓
B	✓	✗	✗	✗	✓
C					
D					
K	✗	✓	✓	✓	✓

The features in Table 2 and others that you can think of are used to make dichotomous keys. There is a key to these species on page 60 of the Unit Guide for F212. You can expect to use such keys in the qualitative tasks in F213 and in F216. Features in dichotomous keys are usually qualitative as they are likely to be specific to certain species.

Try this yourself

1   If you have the Unit Guide to F212 use the dichotomous key on page 60 to identify the ten leaves. If not, there is a key in the Teacher Guidance at:
    http://field-studies-council.org/publications/resources/ks3/key56.htm

2   List some features that are *not* suitable for use in dichotomous keys.

Example 2: leaf shape, types of variable and types of data

Leaves of the holly tree, *Ilex aquifolium*, are often discoloured by a parasitic wasp, called the holly leaf miner, *Phytomiza ilicis*. The adult wasp lays its eggs inside the mesophyll of holly leaves. The eggs hatch into larvae that eat the mesophyll and form 'leaf mines'. A group of students investigated the factors that influence the leaf miner. They made some descriptive observations in note form, drawings and photographs that are summarised in Figure 2. This prompted them to collect some information from 30 leaves. The leaves that they collected formed a **sample** of all the holly leaves in the area. They tried to take samples that were representative of all the leaves. They weighed and measured the leaves, counted the number of prickles and indicated whether each leaf had a mine or not. It was difficult to measure the area of the mines, so they ranked the leaves using a scale of 0 to 10. '0' indicated that there was no mine; '5' indicated that the mine occupied about half of the leaf; '10' indicated that the mine occupies almost the whole leaf.

Figure 2 Some holly leaves with and without mines, showing variation in shape

The data were entered on a spreadsheet. The data collected in an investigation are called the **raw data** to distinguish them from processed data, such as averages. Table 3 shows some of the data collected by one student.

Table 3 Part of a student's spreadsheet with results from sampling holly leaves

Leaf number	Mass/g	Length/ mm	Number of prickles	Presence or absence of leaf mine	Relative area of mine (0–10)
1	0.38	60	16	✗	0
2	0.39	60	19	✓	7
3	0.46	64	15	✗	5
28	1.31	95	5	✗	0
29	1.32	100	17	✓	6
30	1.34	108	7	✓	0

The types of variable and the forms of data are shown in Table 4.

Table 4 Types of variable and types of data

Type of variable	Type of data	Example 1 Holly leaf miner	Other examples in this guide (and page numbers)
Qualitative: categoric	Nominal	Presence or absence of leaf mines	Eye colour of fruit flies (p. 70)
Qualitative: ordered	Ordinal (ranked)	Relative area of each leaf mine	Degrees of cloudiness (p. 25) Abundance scales (e.g. ACFOR) (p. 80)

Type of variable	Type of data	Example 1 Holly leaf miner	Other examples in this guide (and page numbers)
Quantitative: continuous	Interval (having any value, e.g. 1.0, 2.5, etc.)	Length and mass of each holly leaf	Heights of dog whelks (p. 78)
Quantitative: discrete	Interval (integers only, e.g. 1, 2, 3, etc.)	Number of prickles on each holly leaf	Numbers of mollusc species in different areas (p. 75)

Derived variables are those that you use when you have calculated percentage changes or rates of change. They are used for processed data.

Example 3: red blood cells and osmosis

Another simple observational task involves looking at the effect of immersing cells in different concentrations of solutions. In this example, the effect of different dilutions of artificial plasma on the stability of red blood cells is investigated.

The water potential of plasma is approximately the same as that of a 0.9% sodium chloride solution. Percentage solutions are the same as the mass in grams of a compound dissolved in $100\,cm^3$ of water. To make a 0.9% sodium chloride solution, 0.9 g of sodium chloride is added to some water, stirred until it dissolves and then enough water is added to make up the volume to $100\,cm^3$.

Solutions of different concentrations of sodium chloride are prepared. A few drops of blood are added to each solution in a test tube. A bung is placed into the end of the test tube and the tube is inverted to mix the contents. A piece of newspaper is put behind each tube to assess the cloudiness or turbidity of the contents, as shown in Figure 3. Observations are made and recorded. A drop of each mixture is placed onto a microscope slide, a cover slip added and the mixture is observed under a microscope using high power. Further observations are made.

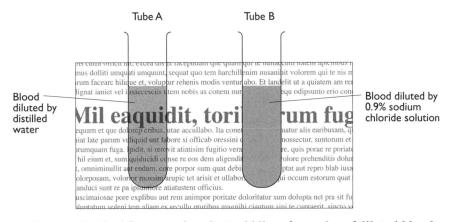

Figure 3 Method for assessing the turbidity of samples of diluted blood

Table 5 Appearance of blood diluted by different concentrations of sodium chloride

Concentration of sodium chloride solution/%	Observations		
	Newsprint legible?	Presence of red blood cells	Appearance of red blood cells
0.0	Yes	No	None present
0.3	Yes	No	None present
0.7	No	Yes	As normal blood
0.9	No	Yes	As normal blood
1.5	No	Yes	Crinkly

It is difficult to observe red blood cells under the microscope, so it is a good idea to think about the theory. For example, you should know that red blood cells will be intact in a solution that has the same water potential as plasma. With 0.9% sodium chloride solution you are effectively adding some blood cells into a larger volume of plasma. The cells should remain intact, so in the sample taken from that solution you should expect to see red blood cells. Adjusting the light and the focusing is critical in finding the red blood cells. Care should be taken to make sure that the stage is not illuminated by too much light and that the high-power lens is as close as possible to the cover slip before focusing by moving the lens *away* from the slide.

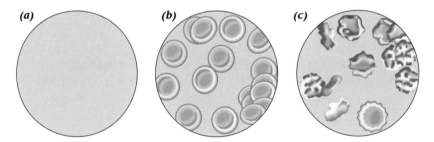

Figure 4 Appearance under the microscope of blood diluted by three different concentrations of sodium chloride solution (a) 0.3%, (b) 0.9% and (c) 1.5%

Knowing that 0.9% sodium chloride solution should contain intact red blood cells you could next look at the two extremes of the concentration range — 0.3% and 1.5%.

As the solution becomes more dilute the water potential gradient between the solution and the red blood cells becomes steeper and water diffuses into the cells by osmosis. This causes them to burst, so you would not expect to see many intact red blood cells in the 0.3% solution. Figure 4 shows that there are none.

Now look carefully at Table 5. You can see that it is possible to read the newsprint through the 0.3% solution, but not through the 0.9% and 1.5%. This is because the blood cells in suspension disperse the light as it travels through the tube. Looking at the microscope slides confirms the presence of red blood cells, although those suspended in 1.5% sodium chloride are misshapen.

In investigations variables are classified as independent, dependent and control:

- The **independent variable** (IV) is the one you have chosen to change (or are told to change in the practical task). In Example 3, the IV is the concentration of sodium chloride. In some investigations there are two or more IVs (see Example 5).
- The **dependent variable** (DV) is the one that you set out to observe and/or measure. There are three DVs in Example 3 — turbidity, presence/absence of red blood cells and their appearance.
- **Control variables** (CV) are those kept constant because otherwise they might have an impact on the values of the DV.

Try this yourself

3 What is the concentration of a 0.9% sodium chloride solution in $g\,100\,cm^{-3}$ and in $g\,dm^{-3}$?

4 When provided with a 2.0% sodium chloride solution, how would you make $10\,cm^3$ of the different concentrations shown in Table 5?

5 Why are the cells in 1.5% sodium chloride solution misshapen?

Example 4: sensitivity of Benedict's test

The range of the independent variable in Example 3 was quite narrow — from 0 to 1.5%. Sometimes you need to use a much wider range, for example one with a difference between concentrations of a factor of 10.

The sensitivity of the Benedict's test is the lowest concentration of reducing sugar that it can detect. To find this, a range of glucose concentrations between 10 and $0.0001\,g\,100\,cm^{-3}$ was prepared by making a serial dilution from the most concentrated solution, as shown in Figure 5.

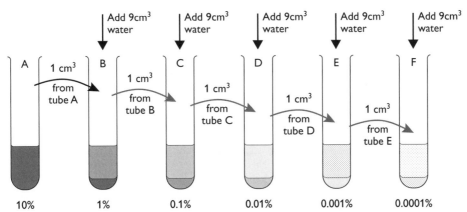

Figure 5 Preparing a serial dilution of glucose

Benedict's solution is added to each solution and the tubes put in a water bath at 80°C for 3 minutes. The colours of the solution are recorded in Table 6.

Table 6 Sensitivity of Benedict's test: to find the lowest concentration of reducing sugar that can be detected

Test tube	Concentration of glucose/g 100 cm^{-3}	Colour after the Benedict's test	
		Student A's results	Student B's results
A	0.0001	Blue	No change
B	0.001	Blue	No change
C	0.01	Blue	Bluey-green
D	0.1	Yellow	Browny-orange
E	1.0	Orange	Reddy-orange
F	10.0	Red	Dark red

You should record observations of qualitative variables such as colour in simple language such as 'blue' or 'orange'. Where fine discrimination is needed, you should use extra terms such as 'pale', 'light' or 'dark', and you may use comparisons such as 'darker red than at 3 minutes' or 'paler green than at 0.2%, but darker than at 0.4%'. You should avoid ambiguous descriptions of colour, such as 'reddy-orange' or 'bluey-green' that Student B has used. 'No change' is not a colour so that should not be used either.

Example 5: inhibition of amylase

This example shows how to record data in tables.

The activity of enzymes is influenced by inhibitors. In an investigation to find inhibitors of the enzyme amylase four substances were tried. A starch suspension was prepared in a buffer solution at pH 6.0. Equal volumes of amylase solution that had been pre-treated with solutions of the inhibitors were mixed with the starch suspension and kept in test tubes at 30°C for 30 minutes. After that time samples were taken and tested in a spotting tile with iodine solution. The results are shown in Figure 6.

Table 7 Tabulated results from Figure 6

Potential inhibitor	Presence (✓) or absence (✗) of starch			
	Source of amylase			
	Bacillus	Aspergillus	Human saliva	Animal pancreas
Copper sulfate	✓	✓	✓	✓
Lead nitrate	✓	✓	✓	✓
Oxalic acid	✓	✓	✓	✓
Silver nitrate	✓	✓	✓	✓
Water	✗	✗	✗	✗

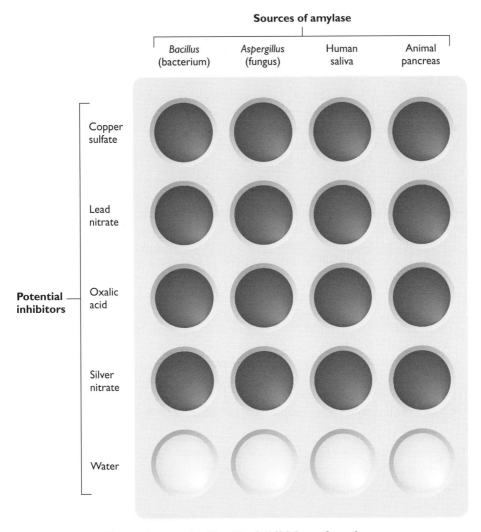

Figure 6 Investigating the inhibition of amylase

Try this yourself

6 In Example 5, what are the independent and dependent variables?

7 Why was water used?

8 What can you conclude from the results in Figure 6 and Table 7?

Before you start to construct a table, decide what you want to record. Decide how many columns and how many rows you will need. Make a rough table in pencil. Make sure you have read all the instructions before you draw the table outline.

Follow these rules:
- Use the space provided; do not make the table too small.
- Leave some space to the right of the table in case you decide you need to add more columns.
- Draw up the table in such a way that you can write observations or readings directly into it, rather than on another page and then copying them into the table (tables should show all the raw data you collect).
- Draw the table outlines in pencil.
- Rule lines between the columns and rows.
- Rule lines around the whole table.
- Write brief, but informative, headings for each column.
- Columns headed with physical quantities should have appropriate SI units.
- When two or more columns are used to present data, the first column should be the independent variable; the second and subsequent columns should contain the dependent variables (sometimes it may be necessary to draw the table in landscape, especially if you are collecting a lot of raw data.
- Rows should be named appropriately.
- Entries in the body of the table should be brief — for example, single words, short descriptive phrases, numbers, ticks or crosses.
- Data should be ordered so that patterns can be seen. It is best to arrange the values of the independent variable in ascending order (see Tables 5 and 6).
- In the quantitative task you will enter numbers into a table. Remember that numbers written into the body of the table do not have units because the units are present in the column headings.
- Wherever possible, tables should be given informative headings.

The solidus or slash (/) meaning 'per' should *not* be used in units. If you have to include concentrations as in Table 6 you do *not* write g per $100 \, cm^3$ as $g/100 \, cm^3$. It should always be written out in full using 'per' or, better, as $g \, 100 \, cm^{-3}$. The negative exponent, as in cm^{-3}, means 'per'.

Note that the solidus is used to separate what is measured from the unit in which it is measured. You may notice that examination papers and some textbooks use brackets around the units in tables. This is also acceptable, but the solidus is the convention used in this unit of study guides for OCR biology.

Table 8 Correct and incorrect ways of showing units in tables and graphs

Correct	Incorrect
Concentration/$g \, 100 \, cm^{-3}$ *or:* Concentration ($g \, 100 \, cm^{-3}$)	Concentration: $g/100 \, cm^3$
Rate of activity/$cm^3 \, s^{-1}$ *or:* Rate of activity ($cm^3 \, s^{-1}$)	Rate: cm^3/s

Qualitative tasks: biological drawings

Another form of qualitative task involves making biological drawings. You are expected to be able to make drawings of prepared sections and also of materials that you have prepared yourself.

Example 6: observing plasmolysis

If you remove a piece of epidermis from an onion scale leaf and mount it on a microscope slide in water you can, using high power, focus on individual cells. You will see the cell wall, the nucleus and the nucleolus. If you then irrigate the tissue with $1 \, mol \, dm^{-3}$ potassium nitrate solution you can observe plasmolysis as the vacuole and cytoplasm lose water by osmosis. Figure 7 shows the appearance of the cells before and after adding the potassium nitrate solution. The diagram includes the label 'space filled with potassium nitrate solution'. You would not be able to see the solution; the annotation is to explain what has happened.

You may be expected to add labels and annotations to your drawings as in Figure 7.

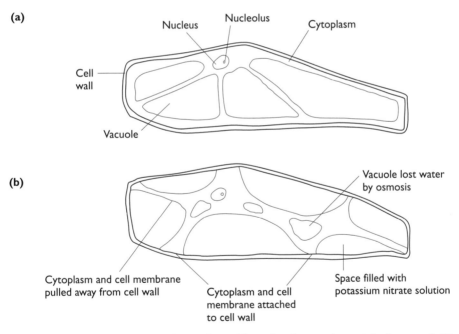

Figure 7 Drawings of a cell from the epidermis of an onion (a) before, and (b) after irrigating with $1 \, mol \, dm^{-3}$ potassium nitrate solution

Sometimes when you look at the epidermis from the storage leaves of onions you can see guard cells and stomata. A similar exercise involves removing the epidermis from an ordinary leaf (not a storage leaf) where there is a much higher density of stomata. You may be asked to mount a piece of epidermis on a slide and draw a pair of guard cells together with surrounding epidermal cells. A similar, but more difficult, task is to make a preparation of cells from the root tip of a plant, such as garlic or onion. These cells are squashed on a slide and stained to show chromosomes in various stages of mitosis. You would then be expected to draw cells in different stages of mitosis. A procedure for this is at:

> **www-saps.plantsci.cam.ac.uk/worksheets/ssheets/ssheet17.htm**

Photographs of the stages of mitosis are at:
> **www.lima.ohio-state.edu/biology/archive/mitosis.html**

Example 7: drawing a tissue map

Figure 8 is a photomicrograph of the cross section of a plant root. You are expected to be able to draw a low-power plan diagram of a plant tissue, such as a plant root.

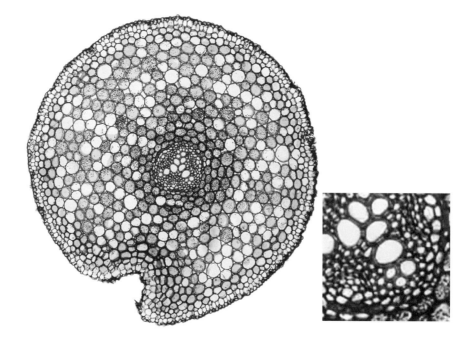

***Figure 8 Photomicrograph of a transverse section of the root of buttercup,
Ranunculus repens. The inset shows part of the central vascular tissue as
seen with the high power of the microscope***

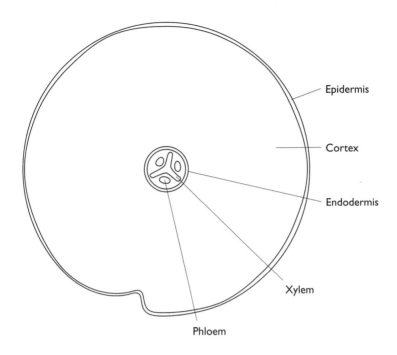

Epidermis

Cortex

Endodermis

Xylem

Phloem

Figure 9 Plan diagram of the transverse section of root shown in Figure 8

Figure 9 is a plan diagram of the root section in Figure 8. When you make a plan diagram, follow these simple rules:

- Make the drawing fill at least half the space provided; leave space around the drawing for labels and annotations.
- Use a sharp HB pencil (never use a pen).
- Use thin, single, unbroken lines (often called 'clear and continuous lines').
- Show the outlines of the tissues.
- Make the proportions of tissues in the diagram the same as in the section.
- Do *not* include drawings of cells.
- Do *not* use any shading or colouring.

Add labels and annotations (notes) to your drawing *only* if they are asked for in the question. Use a pencil and a ruler to draw straight lines from the drawing to your labels and notes. Write labels and notes in pencil in case you make a mistake.

You may also be expected to draw some cells under high power.

Example 8: drawing plant cells

Figure 10 is a drawing of some xylem vessels and phloem tissue from the central vascular tissue in Figure 8.

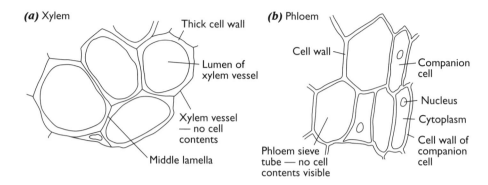

Figure 10 Drawing made under high power of (a) xylem vessels and (b) phloem tissue from Figure 8

High-power drawings should show a small number of cells which should be drawn sufficiently large to show the detail inside. When you make a high-power drawing, follow the same simple rules as for plan drawings, but with the following changes:

- Draw *only* what is asked for in the question, e.g. three cell types or one named cell and all cells adjoining it.
- Show details of the contents of cells such as nuclei and chloroplasts. Draw what you see, not what you know should be present.

General points

In the qualitative task, after the recording of results or the biological drawing, there are a few short questions. You may be asked to carry out a calculation (e.g. magnification or actual size), describe a pattern or trend or give an explanation. This relies on summarising information and using knowledge and understanding of the relevant theory.

You may also be asked to complete a table of comparison. Use only features visible in the biological material you are given. The table outline may be printed on the paper or you may be expected to make the table yourself. If the latter, make sure you include a column headed 'features'.

You may also be asked to explain how the structures you have drawn are related to their functions. For example, xylem vessels are hollow and have no cross walls so they provide little resistance to the flow of water.

Semi-quantitative results

It is possible to make results semi-quantitative as in Example 1 with the area of the leaf mines. These are ranked data that do not have units (see Table 3).

Example 9: loss of chloride ions from plant tissue

Plant tissue, such as the storage tissue of carrots, is cut into thin slices and then washed thoroughly in distilled water. The slices are put into boiling tubes each containing the same volume of distilled water. The boiling tubes are placed in water baths at temperatures ranging from 10°C to 80°C. After 20 minutes, the water surrounding the discs is carefully poured into clean test tubes. The same number of drops of silver nitrate is added to each tube. Water from the tubes that were kept at high temperatures turns cloudy. Water from the tubes kept at low temperature shows no cloudiness. There is a gradation in cloudiness that can be ranked using a scale of 0 to 10; 0 equals no cloudiness and 10 equals most cloudy.

Try this yourself

9 Explain the results of Example 9.

Scales without units are also used in ecology. Examples, such as the ACFOR scale, are described on p. 80.

Example 10: colour standards with Benedict's solution

Ranked data are useful when you do not have access to apparatus such as a colorimeter.

One way to find the approximate concentration of reducing sugar in samples of fruit juice or extracts made from plant tissues is to use a set of colour standards. This is done by making a range of dilutions of glucose and testing them with Benedict's solution. This gives a range of colours. A sample of the extract is then tested with Benedict's solution in exactly the same way and the final colour is compared with the colour standards. The qualitative task in Example 4 uses this method to obtain results.

It is *not* possible to give an exact concentration using this method. It *is* possible to say that the concentration is between two values or is very close to one value. However, this is not an accurate way of determining the concentration of the extracts. The problem is in determining extract concentrations that lie in-between the concentrations of the solutions used to give the colour standards. In Example 4, a wide range of concentrations was used with 10-fold differences in concentration (10% → 1% → 0.1%, etc.). To find the concentration of reducing sugar in a sample it is necessary to narrow the range of concentrations used to make the colour standards. In this example, students recorded the colours and also the degree of cloudiness. Their results are shown in Table 9 and their estimates of the concentration of reducing sugar in three samples are in Table 10.

Table 9 Results of the Benedict's test carried out on a range of concentrations of glucose and three samples with unknown concentrations of reducing sugar

| Concentration of glucose/ g 100 cm⁻³ | Results of Benedict's test | | | |
| | Student A | | Student B | |
	Final colour	Degree of cloudiness	Final colour	Degree of cloudiness
0.1	Green	No cloudiness	Blue	Not cloudy
0.5	Yellow	Slight cloudiness	Blue-green	Slightly cloudy
1.0	Orange	Cloudiness	Light brown	Quite cloudy
2.0	Red	Cloudiness	Dark brown	Very cloudy
4.0	Dark red	Cloudiness	Brick red	Most cloudy
Fruit juice	Orange-red	Very cloudy	Brick red	Very cloudy
Plant extract	Dark green	Cloudy	Dark brown	Cloudy
Body fluid	Green-blue	Slightly cloudy	Green	Not very cloudy

Table 10 Estimates of reducing sugar content of three samples using colour standards as shown in Table 9

| Student | Estimates of the concentration of reducing sugar/g 100 cm⁻³ | | |
	Fruit juice	Plant extract	Body fluid
A	1.5	0.15	0.05
B	4.0	Between 2.0 and 4.0	Between 0.5 and 1.0

Student A has estimated the concentration far too exactly. All it is possible to do is to estimate one of the concentrations used or a range between two, as Student B has done. Both students could have used a scale (0 to 10) to record the degree of cloudiness.

Quantitative tasks

Example 11: decolourising potassium manganate(VII)

Another way of determining the concentration of reducing sugar is to use a solution of potassium manganate(VII). This is reduced to a colourless solution of manganese(II) by reducing agents, such as the reducing sugars glucose and fructose. A small volume of acidified potassium manganate(VII) is added to different concentrations of glucose or fructose. The colour gradually disappears. Timing how long it takes for the pink colour to disappear is another simple way to determine reducing sugar concentration without using a colorimeter.

skills guidance

The problem with timing colour changes is that you can never be certain that you have used the correct end point and that you have used exactly the same end point each time.

Table 11 Results of timing the disappearance of pink colour from potassium manganate(VII) solution after adding different concentrations of glucose

Glucose concentration/ g 100 cm^{-3}	Time for pink colour to disappear/s
6	443
8	289
10	187
15	114
20	57

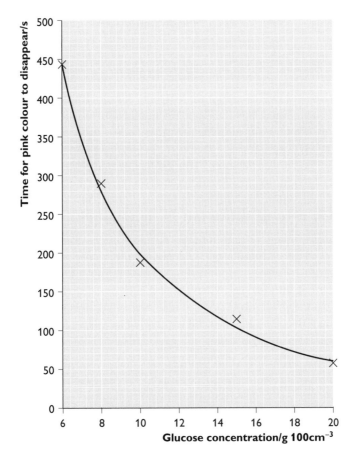

Figure 11 Calibration graph for the results shown in Table 11

To find the reducing sugar concentration in a sample the time taken for that sample to decolourise potassium manganate(VII) solution is measured. Then, by taking an intercept on the graph in Figure 11 the concentration of reducing sugar can be estimated. Drawing a graph makes it possible to estimate intermediate concentrations, rather than giving a range as with the colour standards method (Example 10).

Try this yourself

10 Use Figure 11 to estimate the concentration of solutions that take 90 s and 225 s to decolourise potassium manganate(VII).

Benedict's test and potassium manganate(VII) are not specific for any one type of reducing sugar. Test strips used for testing urine samples contain an enzyme that is specific to glucose.

Two test strips used commonly are Clinistix® and Diastix®. The latter gives a wider range of colours and has a quantitative scale, so is a quick way to obtain semi-quantitative results. Using strips is not very accurate because it relies on matching the colour of the strip to a colour standard provided by the manufacturer.

One way to make colours quantitative is to use a **colorimeter**. A light is shone through the test solution; a light sensor detects how much light passes through the solution and this is expressed numerically. You need to know about colorimeters in Unit F212, but you will probably not use one in a practical task as they are expensive. Results provided in the student sheet in the evaluative task may be from colorimetry. If not, you could suggest using a colorimeter as an improvement.

Example 12: rate of diffusion in agar blocks

Substances diffuse down concentration gradients. It is possible to *notice* that some processes occur faster than others, but to quantify this they have to be timed. For example, agar can be mixed with a pH indicator and an alkali. After solidifying, the agar can be cut into pieces. When these pieces are dropped into $2.0 \, mol \, dm^{-3}$ hydrochloric acid the pH changes and the indicator gradually changes colour as the acid diffuses inwards. Using this method we can tell how long it takes for the acid to diffuse to the centre of an agar block.

Now imagine that these agar blocks are cells. As cells increase in size, the relationship between the surface area and volume changes. By making blocks of different sizes it is possible to investigate the effect of surface area-to-volume (SA:V) ratio on diffusion.

$$SA\text{:}V \text{ ratio} = \frac{\text{surface area}}{\text{volume}}$$

Try this yourself

The table shows some data about acid diffusion into agar blocks of different sizes.

Cube of side/mm	Surface area of cube/ mm²	Volume of cube/ mm³	Surface area: volume ratio	Time for acid to diffuse to centre of cube/s		
				Replicate 1	Replicate 2	Mean
1				8	11	
2				26	26	
3				43	45	
4				65	80	
5				112	122	

11 Copy and complete the table and draw a graph to show the relationship between surface area-to-volume ratio and diffusion.

12 Use your results to explain why multicellular animals need specialised surfaces for exchange of substances with their environment and why they need transport systems.

Model cells can also be made from Visking (dialysis) tubing. It is permeable to water, glucose and salts, but impermeable to larger molecules such as sucrose and starch. Visking tubing is used in investigations on diffusion and osmosis.

Example 13: mass change of potato in sucrose solutions

Example 3 was about osmosis and animal cells. This example is about osmosis and plant cells. As stated in Example 3, red blood cells do not change in volume when they are in 0.9% sodium chloride solution. Immersing plant tissues in solutions of different water potential can be used to determine the water potential of plant tissue. This involves finding the solution in which the plant tissue does not change. This is done by measuring volume or, more often, by measuring mass or length.

Samples of a suitable plant storage tissue such as potato or carrot are cut up into pieces that are weighed and their masses recorded. They are placed in sucrose solutions of different concentrations. After a certain length of time all the pieces are removed from the solutions and dried carefully. They are then reweighed. The results from such an investigation are given in Table 12.

Table 12 The effect of immersion in different concentrations of sucrose solution on the mass of pieces of potato storage tissue

Concentration of sucrose solution/g dm⁻³	Initial mass/g	Final mass/g	Change in mass/g	Percentage change in mass	Mean percentage change in mass
0.0	1.26	1.49	0.23	18.3	19.3
	1.26	1.51	0.25	19.8	
	1.22	1.46	0.24	19.7	
0.2	1.22	1.31	0.09	7.4	5.4
	1.27	1.32	0.05	3.9	
	1.25	1.31	0.06	4.8	
0.4	1.43	1.29	−0.14	−9.8	−7.7
	1.32	1.26	−0.06	−4.5	
	1.37	1.25	−0.12	−8.8	
0.6	1.32	1.10	−0.22	−16.7	−15.2
	1.26	1.07	−0.19	−15.1	
	1.22	1.05	−0.17	−13.9	
0.8	1.28	0.98	−0.30	−23.4	−21.3
	1.25	1.01	−0.24	−19.2	
	1.23	0.97	−0.26	−21.1	
1.0	1.31	0.96	−0.35	−26.7	−23.3
	1.27	1.00	−0.27	−21.3	
	1.23	0.96	−0.27	−22.0	

In this investigation it is necessary to calculate the percentage change in mass because the initial masses of the potato pieces are different. The percentage change in mass is the derived variable. It is calculated as follows:

$$\text{percentage change} = \frac{\text{change in mass}}{\text{original mass}} \times 100$$

In this case, the concentration of the sucrose solution is the independent variable and the change in mass is the dependent variable. Three pieces of tissue were used for each concentration. These are repeats or **replicates** and they were placed in separate test tubes. One piece of tissue is unlikely to be representative of all the tissue of the plant. In practical tasks it is usual to carry out three repeats or replicates for each value of the independent variable.

The balance used for weighing the sections of potato tuber tissue was accurate to the nearest 0.1 g. Some balances are more sensitive and weigh to the nearest 0.01 g. If measuring change in length you should use a ruler that measures to the nearest millimetre. This is a measure of **precision** taking in results. Recording to the nearest

0.1 g is more precise than measuring to the nearest 1 g. Similarly, measuring to the nearest millimetre is more precise than measuring to the nearest centimetre. So 10.6 g is more precise than 11 g and 13 mm is more precise than 1 cm. In this context precision means the number of significant figures or decimal places to which values are expressed.

Try this yourself

13 Why were the individual pieces of potato placed into separate solutions rather than being kept all together in the same beaker?

It is possible to be over-precise. Stop clocks and bench timers may measure to one-hundredth of a second (0.01 s). It is unlikely that you could time a colour change or other event to this degree of precision. Therefore, it is better to express your results to the nearest second (or even to the nearest 15 s or 30 s).

Precision does not only apply to taking results. It also applies to the apparatus used for preparing materials for practical tasks. For example, you may use a syringe for measuring volume. It is not easy to measure precisely with a syringe, particularly with a coloured solution such as potassium manganate(VII) (dark purple) or betalain solution (see AS qualitative task, p.85).

It is possible to improve the precision by using a graduated pipette or a burette, although it is unlikely that you will be given this apparatus in a task. You may be able to write about this when discussing improvements in your evaluation task.

Descriptive statistics

Descriptive statistics are used to summarise the data collected. In Example 2 we looked at the data about holly leaves and the holly leaf miner. The data collected by the student are given in Table 13.

The data on the spreadsheet are organised by increasing length of leaf.

Table 13 Spreadsheet showing the data on holly leaves collected by a student

| | Qualitative variables | | Quantitative variables | | |
| | Categoric | Ordered | Continuous | | Discrete |
Leaf number (x)	Leaf mine	Relative size of mine	Mass/g	Length/ mm	Number of prickles
1	✗	0	0.38	60	16
2	✓	7	0.39	60	19
3	✓	5	0.46	64	15
4	✓	1	0.43	65	16
5	✓	3	0.43	65	15

6	✗	0	0.45	66	14
7	✗	0	0.46	67	22
8	✓	5	0.56	71	20
9	✗	0	0.57	73	20
10	✓	9	0.57	75	19
11	✗	0	0.49	75	17
12	✓	8	0.50	77	12
13	✓	3	0.48	77	19
14	✗	0	0.56	77	16
15	✓	2	0.57	79	17
16	✓	3	0.59	82	19
17	✓	5	0.62	82	19
18	✗	0	0.61	83	2
19	✗	0	0.64	83	7
20	✗	0	0.64	84	19
21	✗	0	0.72	87	19
22	✗	0	0.74	88	5
23	✗	0	0.84	89	20
24	✓	8	0.86	90	17
25	✗	0	0.90	92	19
26	✗	0	0.92	93	6
27	✗	0	0.96	93	17
28	✗	0	1.00	95	5
29	✓	6	1.01	100	17
30	✗	0	1.03	108	7
		Range	1.03−0.38 = 0.65	108−60 = 48	22−2 = 20
		Number (n)	30	30	30
		Mean ($\bar{x}$)	0.65	80	15
		SD	0.20	12	5
		SE	0.04	2	1
		95% CI	0.08	4	2
		Median	0.58	81	17
		Mode	0.57	77	19

Qualitative data

The students were interested in finding out what proportion of leaves was parasitised by holly-leaf miners. In this sample, the percentage of leaves with mines was calculated as follows:

$$\text{percentage of leaves with mines} = \frac{13}{30} \times 100 = 43.3\%$$

Although numbers have been used for recording the relative area of leaf affected by mines, they are ranked data (ordinal data) so '5' does not necessarily represent exactly half of '10'. As a result, it is not possible to calculate the mean. The student could present the qualitative data in columns 2 and 3 as bar charts.

Quantitative data

The student has used the spreadsheet to summarise the quantitative data by showing the following:

- range (difference between largest and smallest)
- mean ($\bar{x}$) calculated as: $\dfrac{\sum x}{n}$

  where $\sum$ = 'the sum of', n = number of readings
- standard deviation (SD or σ)
- standard error (SE or S_M)
- median

The **range** is the difference between the largest result and the smallest. You can express this in two ways, e.g. for the length of leaves as '60 to 108' or as the difference between them, which is 48. The range is one way to show the spread of results about the mean.

The term **average** is often used to mean 'centre of a distribution' which can be the **mean**, the **mode** or the **median**. The student has calculated the **sample mean**. We want to know how this represents the **population mean** — the mean for all holly leaves. We can do this by calculating measures of the spread of results about the mean.

Calculations may generate numbers with many decimal places. Do not write down all the numbers from your calculator display. You must round them up or down and you must do this correctly. You should either round up to the closest whole number or to one or two decimal places to agree with the precision of the raw data.

Variables such as length and mass are likely to be distributed normally so that the mean is in the middle of the range. A graph of continuous data should show a bell-shaped curve indicating a normal distribution. If the data are like this then we can carry out calculations to indicate how much spread there is about the mean.

The **variance** is the difference between each result and the mean. All of these differences are squared to remove the negative signs. The sum of these divided by one less than the number of readings is the variance:

$$\text{variance} = \frac{\Sigma(x - \bar{x})^2}{n - 1}$$

This is used in calculations but rarely used on its own as a descriptive statistic because it is not in the same units as the raw data.

The **standard deviation** is the square root of the variance:

$$\text{standard deviation (SD or } \sigma) = \sqrt{\frac{\Sigma(x - \bar{x})^2}{n - 1}}$$

The standard deviation is used to show how widely the data are dispersed as 68% of results are within one SD of the mean and 95% are within 2 SD of the mean. If results are dispersed widely then you cannot be sure of the value of the mean.

Standard error tells you about the mean of the population from which the sample has come. It is calculated by dividing the standard deviation by the square root of the number of readings:

$$\text{standard error (SE or } S_M) = \frac{\sigma}{\sqrt{n}}$$

This value is used to calculate the **95% confidence interval** (95% CI). There is a 95% probability that the population mean is to be found within the range $\bar{x} \pm \text{CI}$. The 95% CI may be estimated for samples of ≥ 20 by:

$$\bar{x} \pm 2 \times \text{SE}$$

The 95% confidence limits are the upper and lower values of this range. The 95% CI is the best measure of the dispersal of results around the mean and the best to use for error bars on graphs.

It is also the best to use for comparing different sets of data. A line of best fit should go through the error bars. You can expect to be asked to put error bars on graphs and to refer to them in your answers.

For data that are not distributed normally (the graph does not give a bell-shaped curve) the best 'average' to use is the **median**. This divides the data into two equal halves. Quartiles divide the data into quarters. The interquartile range contains all the results within the range 25% below and 25% above the median.

The total range may include the outliers that are discounted because of errors. This means that the total range would give an idea that the real variation is greater than it is. The interquartile range gives most of the data, but with standard deviation and standard error the extent of the deviations from the mean is taken into account.

Error bars can show:
- the total range of results (maximum to minimum)
- interquartile range
- standard deviation

- standard error
- 95% confidence limits

There are several examples of error bars throughout this guide.

Uncertainty in measuring

Uncertainty equates to half the smallest graduation on the apparatus — for example, if the smallest division on a syringe is $1.0\,cm^3$ then the uncertainty is $\pm0.5\,cm^3$. If you start measuring at 0, then the uncertainty applies only to where the measurement is taken, say at $6.3\,cm^3$. The result is expressed as $6.3 \pm 0.5\,cm^3$. However, if a measurement starts at a point other than 0 (for example when taking readings from a burette) then the uncertainty applies to both readings, so it is multiplied by two — for example, $7.5 \pm 1.0\,cm^3$. Similarly, when using a ruler there is an error on each reading unless you start at 0. The same applies to measuring a quantity in a syringe by filling up from empty. The error here is half the minimum measurement. However, when you take two readings from a syringe (say delivering $2.0\,cm^3$ by moving the plunger from $6.5\,cm^3$ to $4.5\,cm^3$) the uncertainty is multiplied by two.

It is possible to calculate the percentage error for a piece of apparatus. Imagine you have collected a gas and measured the volume with a syringe that has graduations every $1\,cm^3$. If you have measured $5\,cm^3$ of gas, you can be certain that you have collected more than $4.5\,cm^3$ but less than $5.5\,cm^3$. The error is $\pm0.5\,cm^3$ in $5\,cm^3$. The percentage error is:

$$\frac{0.5}{5} \times 100 = 10\%$$

If the volume of gas collected is $10\,cm^3$, then the percentage error is 5%.

Accurate data

In Example 4, the students carrying out the investigation to determine the concentrations of reducing sugar in the 'unknown solutions' do not know the true values, but the technician or teacher does. So it is possible to say how accurate the students' results are compared with the true values. In many biological investigations this is not possible; in Example 10, we are only able to say that the concentration of reducing sugar is within a range or is close to a value. The method does not allow a high degree of accuracy. The accuracy can be improved by using a different method. This is done by timing how long it takes for the first appearance of a green colour when the samples are tested with Benedict's solution. This time interval decreases with increasing concentration of reducing sugar. However, this still relies on sight to determine an end point and you may not see the same end point consistently. Another method is to filter the Benedict's solution so that the copper(I) oxide precipitate is removed by the filter paper as a residue. This can be dried and weighed. Using a balance improves the accuracy because it does not rely on taking readings by eye. You could also put the filtrate into a colorimeter and take a reading. Neither of these methods is possible during a practical task, although you may be given such data to evaluate.

Figure 12 shows some hypothetical results. Four students each carried out the same task and each took three replicates. Student **A** has three replicates clustered together near the true value. These results are said to be **reliable** because they are close together. The mean for these results is close to the true value and these results show good accuracy. Student **B** has a good cluster of results but is inaccurate as the mean is a long way from the true value. Student **C** has two results that are close together and one, the outlier, that is far away. Student **C**'s outlier is closer to the true value than the other two readings. Student **D** has results that are widely scattered (not reliable) but the mean will be near the true value.

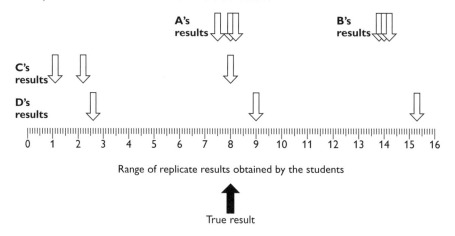

Figure 12 Accuracy: results obtained by four students, A, B, C and D

What should these students do next? In a school or college investigation or in a practical task there is rarely time to repeat the procedure. However, if they all took more results the following might happen:

- Students **C** and **D** may continue to get scattered results, so they should have little confidence in the mean value. Students **A** and **B** may continue to get results that are clustered together and so have mean values about which they are confident. They can both maintain that their results are reliable, but we know that student **B**'s results are inaccurate and student **A**'s are accurate. If the 95% confidence interval is a large number compared with the mean then the data are **unreliable**; if it is small compared with the mean then the data are **reliable**.

The problem in biological investigations is that the true value is not always known. In some cases results can be checked with sources of data. For example, tidal volume readings should be about $500\,cm^3$ and the water potential of blood should be equivalent to 0.9% sodium chloride solution ($-3.86\,MPa$). But the water potential of plant tissues varies considerably and there is no specific value it can be checked against.

This is why, in an investigation, it is important to evaluate the procedure followed and the results obtained. Results, such as the outlier in Figure 12, may be considered anomalous — you need to think about how these might have been obtained. However, what might seem to be anomalies or outliers are not necessarily so, as we can see with student C's results.

Analysis and interpretation of results

Analysing results involves processing raw data and presenting the processed data by drawing a chart or graph. Interpretation involves describing trends and patterns in the data and using knowledge and understanding of the topic involved to make conclusions.

Processing data

Processing data may be as simple as calculating a mean, a percentage or a percentage change. However, it might well involve more. If you have carried out an investigation to find the effect of a variable on the rate of a reaction then the raw data have to be converted into a rate. The results of this processing are derived variables.

Example 14: hydrolysis of starch

The effect of different concentrations of amylase on the rate of starch hydrolysis was determined by finding out how long it took for iodine solution to give a negative result for starch. The experiment was set up as shown in Figure 13 and results were recorded as shown in Tables 14 and 15.

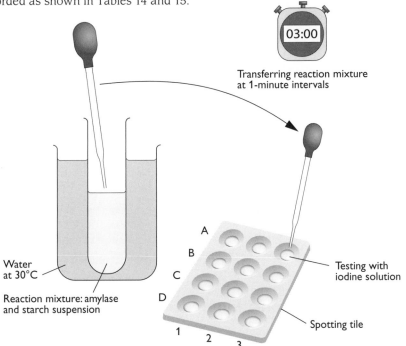

Figure 13 Procedure for investigating hydrolysis of starch by amylase

Table 14 Qualitative data recorded during an investigation of the effect of amylase concentration on the rate of hydrolysis of starch

Time/ min	Colour when iodine solution is added to samples of reaction mixture			
	Concentration of amylase/g dm^{-3}			
	0.0	0.1	0.5	1.0
1	Blue-black	Blue-black	Blue-black	Very dark brown
2	Blue-black	Blue-black	Very dark brown	Dark brown
3	Blue-black	Dark brown	Dark brown	Dark brown
4	Blue-black	Dark brown	Light brown	Light brown
5	Blue-black	Light brown	Light brown	Yellow (no change)
6	Blue-black	Light brown	Light brown	Yellow
7	Blue-black	Light brown	Yellow (no change)	Yellow
8	Blue-black	Light brown	Yellow	Yellow
9	Blue-black	Yellow (no change)	Yellow	Yellow
10	Blue-black	Yellow	Yellow	Yellow

The hydrolysis of the starch is complete when there is no change in the colour of the iodine solution.

Table 15 Rates (derived from Table 14) of hydrolysis of starch by different concentrations of amylase

Concentration of amylase/g dm^{-3}	Relative rate of activity/ min^{-1}
0.0	0.0
0.1	1.1
0.5	1.4
1.0	2.0

In this example the time taken for a process is recorded. Since a fast reaction occurs in a short period of time, the short times need to be turned into fast rates. To do this you need to calculate the reciprocal, $1/t$, where t is the time taken. However, this often gives very small numbers, so it is better to use 10, 100 or 1000 as the numerator. In Example 14, 10 has been used. The reciprocal is an estimate of the overall rate. Be aware that during the reaction the rate changes as the concentration of substrate decreases. Do not put the calculations into your tables. Only write the final result, as shown in Table 15.

Example 15: decomposition of hydrogen peroxide

The enzyme catalase is distributed widely in animal and plant tissues. It catalyses the decomposition of hydrogen peroxide, forming oxygen as a product. The reaction rate can be calculated by working out how much product is produced per unit of time. In enzyme investigations the substrate is used up, so the concentration of substrate decreases and the rate also decreases. Therefore, it is usual to calculate the initial rate. This is done by taking measurements over a short period of time, plotting them on a graph and drawing a tangent in order to calculate the rate.

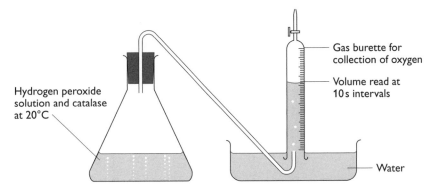

Figure 14 Procedure for investigating the time course of the decomposition of hydrogen peroxide by catalase

The experimental procedure is shown in Figure 14. Some students' results are shown in Table 16. To calculate the initial rate of reaction, plot the figures for each concentration of hydrogen peroxide and draw a tangent as shown in Figure 15. The points that you use to calculate the rate should be separated by at least half the length of the tangent.

Table 16 Volumes of gas collected at 10s intervals during the decomposition of five different concentrations of hydrogen peroxide catalysed by catalase (raw data)

	Volume of gas collected/cm^3				
	Concentration of H$_2$O$_2$/mol dm^{-3}				
Time/s	0.2	0.4	0.6	0.8	1.0
0	0.0	0.0	0.0	0.0	0.0
10	0.7	1.3	1.8	2.1	2.4
20	1.4	2.1	2.6	3.2	3.8
30	1.9	2.8	3.2	3.9	4.7
40	2.3	3.2	3.6	4.4	5.3
50	2.8	3.6	4.1	4.8	5.6
60	3.2	4.0	4.6	5.2	5.7
70	3.6	4.4	5.0	5.4	5.8
80	4.0	4.7	5.4	5.7	5.9
90	4.4	5.2	5.7	5.9	6.0
100	4.6	5.5	6.0	6.0	6.0

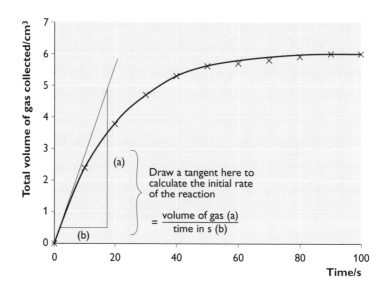

Figure 15 Calculating the initial rate of an enzyme-catalysed reaction

Table 17 The effect of concentration of hydrogen peroxide on the rate of its decomposition by catalase (processed data)

Concentration of H_2O_2/mol dm^{-3}	Rate of reaction/cm^{3}s^{-1}	Rate of reaction x 10/cm^{3}s^{-1}
0.2	0.07	0.7
0.4	0.13	1.3
0.6	0.18	1.8
0.8	0.23	2.3
1.0	0.27	2.7

Note that the rates have been multiplied by ten on the third column to make them easier to plot on a graph.

Try this yourself

14 Plot the results shown in Table 17 on a graph.

15 Explain the effect of increasing the concentration of the substrate on the initial rate of the reaction.

Example 16: effect of pH on amylase

The experimental apparatus shown in Figure 13 can be used to investigate the effect of pH on the activity of amylase. Tables 18 and 19 show some experimental results.

Table 18 The effect of pH on the rate of starch hydrolysis by amylase (raw data)

pH	Time taken for no colour change with iodine solution/s		
	Replicate 1	Replicate 2	Replicate 3
3	100	90	100
5	15	15	15
7	30	20	30
9	120	120	140
11	490	460	470

Table 19 The effect of pH on the rate of starch hydrolysis by amylase (processed data)

pH	Mean time taken for no colour change with iodine solution/s	Rate of reaction/s^{-1}
3	96.7	10.3
5	15.0	66.7
7	26.7	37.5
9	126.7	7.9
11	473.3	2.1

The results in Table 19 are plotted on Figure 18 (p. 44).

Example 17: catalase in various plant tissues

The apparatus used in Figure 14 can be used to investigate the activity of catalase in different plant tissues. Some student results are shown in Table 20.

Table 20 Rate of decomposition of hydrogen peroxide by catalase in six different plant tissues

Plant material	Time taken to collect 5 cm^3 gas/s	Rate of gas production/ cm^3 s^{-1}
Celery	11	0.455
Carrot	132	0.038
Potato	23	0.217
Lettuce	51	0.098
Mung beans	4	1.250
Apple	255	0.019

Presenting processed data

Tabulated results such as those in Example 15 may reveal a trend or pattern. However, it is not always easy to see a trend or pattern in tabulated data so it is necessary to draw a graph. There is unlikely to be enough time to draw a graph during Part 1 of the quantitative task, so it is likely to be in Part 2. You may also have to draw a chart or graph of given data in the evaluative task. You need to be confident about drawing charts and graphs because it takes time to choose the appropriate way to present the data, rule axes, decide on scales, plot the points and then decide what sort of line to draw.

In Example 17, the independent variable is qualitative as it is discrete categories and is not continuous. In some investigations you may have variables like this (see Example 24, p. 63). A bar chart (see Figure 16) is the appropriate way to present these results.

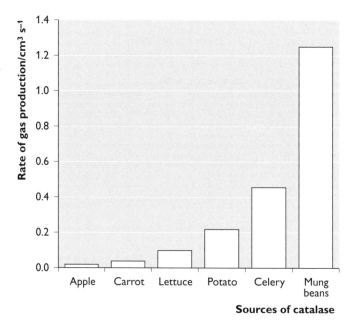

Figure 16 A bar chart of the results from Example 17: catalase activity in various tissues

Rules for drawing bar charts are as follows:
- Use at least half the grid provided; do not make the chart too small.
- Draw the chart in pencil.
- Bar charts can consist of lines, or blocks of equal width, that do not touch.
- The intervals between the blocks on the x-axis should be identical.
- The y-axis should be properly scaled with equal intervals.
- The y-axis should be labelled with units.
- The lines or blocks can be arranged in any order, but to make comparisons it can help if they are arranged in descending/ascending order of size.

- Each block should be identified. There is no need to shade the blocks or colour code them.

Do not confuse bar charts with histograms. A histogram is drawn for continuous data that are subdivided into classes. A good example is collecting data on continuous variables, such as linear measurements or mass. If you are analysing data then you may wish to draw a frequency histogram to see if the data show a normal distribution. Figure 17 shows a histogram for the data on holly leaves from Table 13.

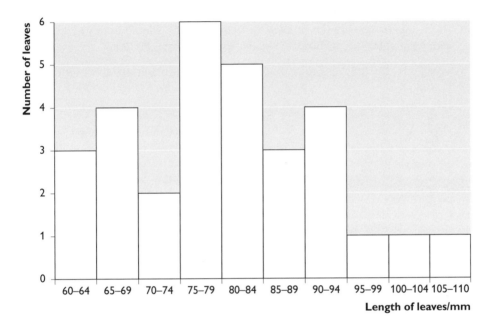

Figure 17 Histogram of leaf length for the data on holly leaves in Table 13

Histograms are used when the independent variable is numerical and the data are continuous. They are sometimes referred to as frequency diagrams.

First, the raw data have to be organised into classes:
- The number of classes has to be established. This will depend largely on the type and nature of the data.
- The range within each class has to be determined. This is usually the total range divided by one less than the number of classes.
- There should be no overlap in the classes, for example:
  4.01 to 5.20 *or* 4.01 < 5.21
  5.21 to 6.40 *or* 5.21 < 6.41

The data should be organised using a tally chart and drawing 'five-bar gates' (see p. 70).

Follow these rules when drawing a histogram:
- Use at least half the grid provided; do not make the histogram too small.

- Draw the histogram in pencil.
- The *x*-axis represents the independent variable and is continuous. It should be properly scaled and labelled with appropriate units.
- The blocks should be drawn touching.
- The *area* of each block is proportional to the size of the class. It is usual to have similar-sized classes so the widths of the blocks are the same.
- The blocks should be labelled either by putting the class ranges (e.g. 60–64, 65–69 etc.) underneath each block or by putting the lowest number in each range (e.g. 60, 65, 70, etc.) under the left-hand side of the relevant block.
- The *y*-axis represents the number or frequency and should be properly scaled with equal intervals. It should be labelled with appropriate units.

In Example 16, the independent variable is continuous so a line graph (Figure 18) is appropriate.

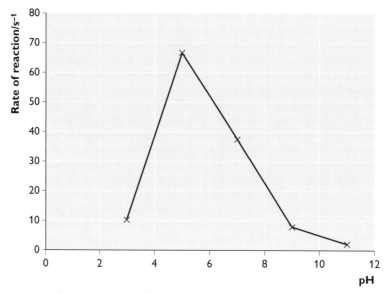

Figure 18 The effect of pH on the activity of amylase

Line graphs are used to show relationships in data that are not immediately apparent from tables. The term 'graph' applies to the whole representation. The term 'curve' should be used to describe both curves and lines used to join points.

The following guidelines should be followed:
- Use at least half the grid provided; do not make the graph too small.
- Draw the graph in pencil.
- The independent variable should be plotted on the *x*-axis.
- The dependent variable should be plotted on the *y*-axis.
- Each axis should be marked with an appropriate scale. The data should be examined critically to establish whether it is necessary to start the scale(s) at zero. If not, you may have a displaced origin for one or both axes (see Figure 11, p. 27).

- Each axis should be scaled using multiples of 1, 2, 5 or 10 for each 20 mm square on the grid. This makes it easy for you to plot and extract data. Never use multiples of three.
- Each axis should be labelled clearly with the quantity and SI unit(s) or derived units as appropriate, e.g. time/s and concentration/g dm^{-3}.
- Plotted points must be marked clearly and be easily distinguishable from the grid lines on the graph. Encircled dots or saltire crosses (X) should be used; dots on their own should not. If you need to plot three lines, vertical crosses (+) can also be used.

After plotting the points you need to decide if any are anomalous. Ask yourself the question: 'do they fit the trend?' You should know something about the theory behind the investigation so you should be aware of the likely trend. If you think one (or more) of the results is anomalous, then it is a good idea to ring it (or them). Draw a circle on the graph away from the line and put a key to state that the circled point(s) represent anomalous result(s). The next thing to decide is how to present the line.

- It may be obvious that the points lie on a straight line. First, decide whether to include the origin (0, 0) if it is not a datum point in your results. If including the origin, place a clear plastic ruler on the grid and draw a straight line from the origin making sure that there is an even number of points on each side of the line. If the origin is not a point, or you are not sure, then start the line at the first plotted point. Do not continue the line past the last plotted point.
- You should only draw a smooth curve if you *know* that the intermediate values fall on the curve. Decide whether the origin is a point and if not start at the first plotted point. The curve should go through as many points as possible, but try to make sure there is an even number of points on each side of the line. Do not continue past the last plotted point.
- If you are not sure whether the relationship is a straight line or a curve, draw straight lines between the points. This indicates uncertainty about the results for values of the independent variable between those plotted.
- If a graph shows more than one line or curve, then each should be labelled to show what it represents.

Bar charts, histograms and line graphs should have informative titles.

If the data include times in seconds, do not use minutes as the unit on a graph. It is difficult to use a scale with each small square representing 3 or 6 seconds. Unless the unit for time is whole minutes (see Example 14) always plot in seconds.

Sometimes the variables cover such a wide range that it is impossible to use a normal scale. This is when a log scale may be used — the intervals on the axis increase by an order of magnitude (×10). This is appropriate when a wide range has been used, as in Example 4 with glucose concentrations between 0.0001% and 10%. Log scales are also useful for plotting data on the growth of microorganisms. If intervening numbers have to be plotted then special log graph paper is needed, as a log scale is not divided equally (see Figure 19).

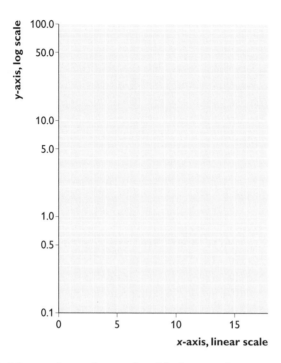

Figure 19 A log scale on the y-axis with three cycles: 0.1 to 1.0, 1.0 to 10.0 and 10.0 to 100.0

As you draw your graph, think about how you are going to describe what it shows. You may be required to describe the trend or pattern, for which you need a vocabulary of words such as:

- decrease
- increase
- peak
- maximum
- optimum
- directly proportional
- indirectly proportional
- constant
- fluctuate

Make sure that you quote data from the graph in support of your description.

Using graphs

There are various ways in which you might be expected to use a graph to extract data, for example:

- find an intermediate value by using an intercept
- predict what would happen beyond the range of the independent variable investigated by continuing the line (extrapolation)

- calculate a rate by using a tangent
- calculate a rate by using a gradient

The processed data in Table 12 (p. 30) can be plotted to show the effect of immersing potato storage tissue in sucrose solutions of different concentrations. In some solutions the mass of the potato cores increases, in others it decreases. If a curve is drawn through the points, it crosses the line at 0% where there would be no change in mass. This intercept represents the water potential of the surrounding solution at which there is no net movement of water between the tissue and the sucrose solution. The water potential of the tissue is the same as the water potential of the solution.

Try this yourself

Water potentials are expressed in pressure units.

16 Plot the data from Table 12 to find the water potential of the potato tissue. The water potentials of the sucrose solutions are given in the following table.

Concentration of sucrose solution/$mol\,dm^{-3}$	Water potential/kPa
0.20	−540
0.40	−1120
0.60	−1800
0.80	−2580
1.00	−3500

Implementing

If you find practical work difficult, then here are some simple pieces of advice that may help you.

Following instructions

In the qualitative and quantitative tasks you follow instructions. Before you start the practical work read these to the end. You may annotate the instructions if you wish. This will help you to understand what you are expected to do during the practical. It will tell you what sort of results to record. The instructions are written so you can follow them in sequence. Read the first instruction carefully and then carry it out. It is a good idea to put a tick by each instruction when you have completed it. Proceed carefully through the rest of the instructions, double checking that you are doing the right thing.

Safety

You are expected to work safely, so follow these rules:
- Keep your work area well organised and tidy.
- Use one area for wet work and keep another area dry for writing on your exam paper; do not do the practical work over the paper.
- Wear protective goggles/spectacles when using liquids or when cutting anything.
- Make sure that you are familiar with the standard hazard warning symbols.
- Take care when using knives, scalpels, glassware, chemicals, Bunsen flames, hot water, etc.
- Inform the teacher or technician immediately if you have an accident.
- Clear up any spillages immediately.

Your teacher will assess your approach to working safely. One of the marks for the qualitative task is for working safely.

Table 21 shows examples of poor laboratory practice that teachers will use as reasons for not awarding marks for safe working and skilful practice. It also shows equivalent good practice.

Table 21 Examples of poor and good laboratory practice

Poor practice	Good practice
Using incorrect apparatus or equipment without realising*, e.g. using a $10\,cm^3$ syringe instead of a $1\,cm^3$ one to dispense $0.5\,cm^3$ of a liquid	Choosing the correct apparatus/equipment/chemicals from those provided so that the correct results are obtained
Using the same syringe to dispense different liquids without realising* so that contamination occurs	Using separate syringes when necessary and keeping syringes separate once used (for correct re-use) *or* washing out a syringe before using it to dispense a different solution
Cutting slices, cubes or sections carelessly so that slices or sections are of uneven thickness, and cubes are of unequal size	Using cutting equipment (e.g. scalpels, knives) and measuring equipment (e.g. rulers) with care to produce slices and cubes of correct size
Allowing fluids to drip off the outside of beakers /tubing/stirring rods/tissue samples into other solutions so that there is the risk of cross-contamination	Rinsing and drying equipment when necessary, clearing off spills from the outside of beakers; keeping different items in clearly defined areas on the bench
Haphazard use of the stop clock/bench timer so that incorrect times are recorded; samples are not taken at correct time intervals	Checking how to use the stop clock/bench timer before you start; careful checking of times; re-setting to zero when required

Poor practice	Good practice
Filtering suspensions through a filter funnel where the filter paper has not been folded correctly/has a tear/has not been fitted into the funnel correctly	Folding a piece of filter paper and then opening it out into the filter funnel; filtering suspensions so that a clear solution, the filtrate, runs through and the precipitate/larger insoluble particles remain on the filter paper

*Realising a mistake and asking for fresh syringes is considered good practice

Think about limitations

As you carry out the quantitative task think about the limitations of what you are doing. Highlight on the question paper the aspects of the procedure that you found difficult or that you had to take extra care over. It is likely that these are possible limitations that you can identify.

Work confidently

Take care over setting up your apparatus, measuring out volumes and taking results. Your teacher will assess how skilfully you work as this contributes to the mark for the tasks.

Bear in mind that you will not achieve the outcomes expected by the examiners unless you take care over such things as, for example, measuring volumes of solutions.

Evaluative tasks

When evaluating an experimental procedure it is important to consider the way in which the procedure was carried out and the quality of the data collected. You need to ask yourself the question: 'can I have confidence in my data?' If you do not have confidence in the data then you cannot have confidence in the conclusion(s) you make. Questions are often asked in these terms.

Reliability is a measure of the closeness of agreement between individual results. It is a measure of the 'repeatability' of the data collection process. However, as shown in Figure 12, closeness of replicates does not mean that the data are accurate. **Reliable results** are repeatable and reproducible. **Repeatable results** are replicate results that are in close agreement. You can use mathematical methods to describe the variation in replicate results (see pp. 33–34). **Reproducible results** are results that can be reproduced by someone else following exactly the same procedure. You can only comment on this in the evaluative task if you are given results from other students.

Precision refers to the number of significant figures (or decimal places) in readings.

Accuracy is a measure of the closeness of agreement between individual results or a set of results and an accepted 'true' value. In biology it is often difficult to know the 'true' value.

Anomalous results are results that do not fit the trend. They are sometimes known as outliers. They can be:
- replicate results that differ significantly from others
- a result (which may or may not be a mean) for one value of the independent variable that does not fit the overall trend and is not included in the curve of best fit

You may be asked to suggest likely explanations for anomalous results. The anomaly could be the first result taken before the experimenter was confident in the procedure.

Validity refers to the confidence that you can have in your conclusions. In a valid investigation:
- you have measured what you set out to measure
- changing the independent variable led to changes in the dependent variable which you have measured

If asked to comment on the validity of an investigation then you should consider the following:
- the limitations in the procedure
- any uncontrolled variables
- the effects of errors (systematic and random) on the results
- the reliability and repeatability of the results
- the precision of the data collected
- the accuracy of the results

Give some positive aspects of the investigation first, followed by some criticisms. You should refer to specific aspects of the procedure and results, rather than using vague comments such as 'my conclusion is valid because my results are precise, reliable and accurate' — this is meaningless without supporting information. For example, you could say that your results are precise because you measured to two decimal places and that they are reliable because the replicates are close together and there are no anomalous data. Your results may be accurate because they agree with an expected trend. Always quote some examples of your raw or processed data in support of your arguments.

The first thing to do when evaluating is to consider the procedure followed. Is it possible that there were any measurement errors in the method? There are two types of error:
- **Systematic errors** are the same throughout the investigation. A common systematic error is that the measuring device may give readings that are out by a certain value. It could be that one of the controlled variables is always incorrect by the same quantity. If there are small systematic errors then the data may be precise, but not accurate. The effect is to overestimate or underestimate the true values.

- **Random errors** occur when you do not carry out the procedure in exactly the same way each time. You may also read the apparatus in a slightly different way. These errors affect some of the results, but not all of them. They do not always affect the results in the same way. Random errors could also be the result of variation in biological material.

Do not think of errors as mistakes. Even in a perfectly performed investigation there will be errors. Systematic errors may not be easy to identify, but you should always check the accuracy of measuring instruments such as balances, colorimeters and pH meters. Random errors should show up in the data, making the data less reliable. However, random errors may affect one value of the independent variable, but not all (see Figure 20).

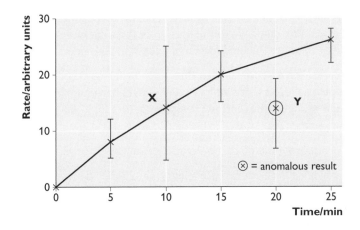

Figure 20 The effect of random errors on processed data

The results plotted in Figure 20 are mean values with range bars. We can see that the result labelled **X** has a much wider range than the others. This suggests that there may have been one, or more, random errors in the readings for 10 minutes. The result labelled **Y** does not fit the overall trend of the other results. This may be because there were random errors when collecting all the data for 20 minutes.

You should consider the variables involved in the investigation. In Example 3 errors of both types may affect the way the sodium chloride solutions are prepared. This would influence the accuracy of the values of the independent variable. In Example 13, errors of both types may influence the results collected. In this case it is the values of the dependent variable that are inaccurate. Controlled variables are the other variables that you were not investigating. Sometimes the instructions will tell you to keep these constant. In Example 15, the temperature may be kept constant by placing the reaction mixtures in a thermostatically controlled water bath at, say, 20°C. The pH may also be controlled by using a buffer solution of, say, pH 7.0. If variables are not controlled then they may influence the results; they are called confounding variables or uncontrolled variables. Sometimes, as in field studies, it may only be possible to be aware of such variables and then 'take them into account' when analysing and interpreting data.

Controlled variables should not be confused with a control experiment. It is important to know that your results are valid, that they show what you think you are investigating. It is necessary to rule out other possible causes. Example 18 includes some control experiments so that you can see why they are important.

Example 18: catalase and hydrogen peroxide

The effect of pH on the rate of an enzyme-catalysed reaction can be studied using cylinders of potato. The cylinders are cut into discs and placed into specimen tubes containing buffer solutions at different pH values. After 5 minutes in the specimen tubes, each potato disc is removed and placed in a specimen tube containing hydrogen peroxide. The disc falls to the bottom of the tube and catalase molecules in the potato disc catalyse the breakdown of hydrogen peroxide.

Bubbles of oxygen form in the potato disc and on its surface. These reduce the density of the disc so it rises. The rate of reaction is determined by timing how long it takes between dropping the disc and it reaching the top of the hydrogen peroxide in the specimen tube.

A number of controls should be carried out. They answer the following questions:

(1) Hydrogen peroxide decomposes in the light. How much decomposition occurs in the length of time the procedure takes?

(2) Hydrogen peroxide decomposes on surfaces. How much decomposition happens on or inside the disc that is not catalysed by the enzyme? This is sometimes called the non-enzymic reaction.

(3) Are there any other molecules in the potato disc that might catalyse the reaction?

The experimenter could devise the following control experiments to answer these questions:
- Place the same volume of hydrogen peroxide into a specimen tube and see if there is any decomposition. If so, then there is some decomposition occurring in the light.
- Boil some cylinders of potato to denature enzymes and then cut into discs. Put these discs into hydrogen peroxide. If any gas is produced then there is some non-enzymic breakdown of hydrogen peroxide.

If an investigation includes one or more controls and the likely confounding variables are controlled then there is a good chance that the data collected are valid.

In the evaluative task you can expect to be asked to identify limitations of the procedure. As you carried out the quantitative task you should have identified some potential limitations. Some of these are likely to occur in almost all laboratory tasks; some are more likely to be specific to the particular task.

It is possible that you were not confident about taking the first few results. You may not have had time to repeat these early results, so they are likely to be inaccurate. Timing is always a problem. Working on your own, it is not possible, for example, to start a stopwatch or bench timer at exactly the same time that you mix two solutions or a potato disc hits the surface of hydrogen peroxide. If you think you were always 5 seconds too slow in starting the timer then you have a systematic error. If your timing method improved then you have a random error that is more significant in the early readings.

Think about the effect of the limitations on the data. You may be asked about this. For example, in Example 18 if you start the timer too early then the length of time recorded will be longer than it should be. If you start it too late, the length of time recorded will be too short. In the first case you will overestimate the length of time; in the second case you will underestimate the time. If you then calculate rates of reaction, then the longer time will give you an underestimate of the rate; a shorter time will give you an overestimate of the rate.

In Example 13, the potato sections are removed and blotted dry before weighing. This surface drying may be more effective on some sections than others. If more solution is left on some sections, then the potato will weigh more than it should.

Try this yourself

17 What would be the effect on the data of not drying the potato sections effectively?

18 How would this affect the determination of the water potential of the tissue?

Limitations and improvements

In this guide it is only possible to identify a few common limitations.

Having identified some limitations and explained the effects on the data you are expected to give an improvement for each limitation. These should be made in the context of the investigation you have carried out. You should not devise a totally new procedure. There are some standard improvements to some of the limitations (see Table 22). In Table 22 the numbers in brackets refer to examples in this guide.

Investigating the effect of pH on enzyme activity may not produce valid results if carried out without using a thermostatically controlled water bath (limitation) as the temperature of the laboratory may fluctuate (temperature not controlled) and enzyme activity changes with temperature (explanation). You would repeat the investigation using a thermostatically controlled water bath set to the same temperature for all replicates (modification/improvement).

Table 22 *Common limitations of practical procedures, their effects on data and suitable improvements)*

Limitation	How the limitation may influence the procedure or results	Example of modification/ improvement to overcome limitation
One measurement or only two replicates (5, 12)	Unable to detect anomalous results thus decreasing reliability	Perform at least three replicates
Difficulty in judging colours or looking for colour changes when using 'by eye' methods (4)	Lack of consistency in judgement may lead to less precise, less reliable and less accurate results	Use colour standards for comparison; use a colorimeter
Insufficient intermediate readings taken, e.g. every 60 seconds (14)	Lack of results between the intervals — could miss changes so making conclusions about trends less valid	Take readings/measurements at smaller intervals, e.g. at 30 s not at 60 s
Insufficient number of values of independent variable, e.g. only at 20°C, 30°C, 40°C, 80°C, 100°C	As above	Include more intermediate temperatures within the range, e.g. 25°C, 35°C, 45°C, 50°C, etc.
Insufficient range of independent variable, e.g. 30°C to 50°C	Lack of results beyond the range investigated — could make it difficult to identify trend or pattern	Extend the range, e.g. 10°C to 80°C
Inconsistent stirring of solutions, e.g. reaction mixture not stirred same way before taking samples (15)	Inconsistent rates of product formation	Standardise timing and method of stirring
Samples taken from different sources, e.g. potato cylinders from different potatoes (14)	Different age/state may introduce random errors	Use the same tissue source e.g. you may have to take smaller cylinders but all from the same potato
Method of timing, e.g. stop clock (17)	Used to a greater level of precision (e.g. 0.01 s) than sensible	Record times to nearest second or nearest 10 seconds, as appropriate
Impossible to start all the reactions at the same time (18)	Some reactions occurred for longer (or shorter) than timed	Use a staggered start, e.g. start the reactions at 1-minute intervals
pH not controlled, e.g. in an enzyme experiment (15)	pH affects enzyme activity	Use a buffer solution to control pH

F216: Practical work at A2

The format for the practical tasks in Unit F216 is exactly the same as for F213 — a qualitative task, a quantitative task and an evaluative task. As the tasks are essentially the same this part of the guide deals with the topics from Units F214 and F215 that are most likely to provide the contexts for the practical assessment in F216.

There are extra requirements for the practical tasks at A2:
- There is a synoptic assessment, in that topics from the AS units may also be assessed.
- Data analysis may be more demanding as it may involve a statistical test.

In Table 23, the AS units in brackets provide the synoptic elements of the examples.

Table 23 Topics from A2 Units F214 and F215 used for examples of practical tasks

Example		Unit	Modules	Task	Skill
19	Chloroplasts and DCPIP	F214	4.3.1	Qualitative	(b)
20	Limiting factors and photosynthesis	F214	4.3.1	Quantitative	(b) (c)
21	Leaf discs and photosynthesis	F214	4.3.1	Quantitative Evaluative	(b) (c) (a) (d)
22	Vital stains and respiration	F214 (F211)	4.4.1 (1.1.2)	Qualitative	(b)
23	Determining rates of respiration	F214	4.4.1	Quantitative	(b)
24	Respiration and respiratory substrates	F214 (F212)	4.4.1 (2.1.1)	Quantitative	(b)
25	Effect of temperature on respiration	F214 (F212)	4.4.1 (2.1.3)	Quantitative Evaluative	(b) (c) (a) (b) (c)
26	Photosynthesis and respiration	F214	4.3.1 and 4.4.1	Evaluative	(a)
27	Problem solving — excretion	F214	4.2.1	Qualitative	(b)
28	High-power drawing of the pancreas	F214	4.1.3	Qualitative	(b)
29	Products of meiosis	F215	5.1.2	Qualitative	(b)
30	Discontinuous variation	F215	5.1.2	Quantitative	(b) (c)
31	Immobilised enzymes and cells	F215 (F212)	5.2.2 (2.1.3)	Qualitative	(b)
32	Kinesis	F215	5.4.3	Quantitative	(b) (c)
33	Taxis	F215	5.4.3	Quantitative	(b) (c)

Example		Unit	Modules	Task	Skill
34	Random sampling	F215 (F212)	5.3.1 (2.3.1)	Qualitative Quantitative	(b) (b) (c)
35	Association between species	F215	5.3.1	Quantitative	(b) (c)
36	Line transects and abiotic factors	F215 (F212)	5.3.1 (2.3.1)	Qualitative	(b)
37	Continuous variation	F215	5.1.2	Quantitative	(b)
38	Belt transects	F215 (F212)	5.3.1 (2.3.1)	Quantitative Evaluative	(b) (d)

Photosynthesis

Example 19 (qualitative): chloroplasts and DCPIP

A task might involve investigating the effects of different wavelengths of light on a suspension of chloroplasts. Leaves of plants such as lettuce or spinach are homogenised with sucrose solution in a blender. The homogenised leaf material is filtered to remove xylem vessels, cell walls and tissues that have not been broken down. The filtrate consists of cell contents, especially chloroplasts. This dark green liquid is kept cold in the dark until required. The action of the light-dependent stage of photosynthesis is investigated by mixing the chloroplast extract with dichlorophenolindophenol (DCPIP), a blue dye that is decolourised when reduced. DCPIP is an example of a redox dye.

To investigate the effect of different wavelengths of light the chloroplast extract is first mixed with DCPIP solution and the resultant blue mixture is placed in test tubes. When a light is shone on the tube the blue colour disappears gradually, leaving green. This shows that DCPIP has been reduced and that the light-dependent stage has occurred. Coloured filters that transmit different wavelengths of light can be placed in front of the tubes. A tube can also be placed in the dark by wrapping it in aluminium foil. A sample of the extract can be boiled and then added to DCPIP. It is a good idea to have a tube without DCPIP and another tube which has only DCPIP. The results and deductions are shown in Table 24.

Table 24 Results and deductions from an investigation of the light-dependent stage of photosynthesis using a redox dye, DCPIP

	Contents					
Tube	Chloroplast extract	DCPIP	Filter	Initial colour	Final colour	Deduction
A	✓	✗	None	Green	Green	–
B	✗	✓	None	Blue	Blue	–
C	✓	✓	None	Blue	Green	DCPIP reduced in white light

Tube	Contents		DCPIP	Filter	Initial colour	Final colour	Deduction
	Chloroplast extract						
D	✓		✓	Red	Blue	Green	DCPIP reduced in red light
E	✓		✓	Blue	Blue	Green	DCPIP reduced in blue light
F	✓		✓	Green	Blue	Pale blue	DCPIP reduced slowly in green light
G	✓ Boiled		✓	None	Blue	Blue	Reduction of DCPIP requires functioning proteins in chloroplasts
H	✓ Treated with copper		✓	None	Blue	Blue	Copper ions inhibit photosynthesis

Try this yourself

19 Why are tubes A and B included?

20 Explain how quantitative results could be obtained from this investigation.

Example 20 (quantitative): limiting factors and photosynthesis

The apparatus shown in Figure 21 is used to investigate the effect of limiting factors (light intensity, carbon dioxide concentration and temperature) on photosynthesis in aquatic plants, such as fanwort, *Cabomba spp.*, and Canadian pondweed, *Elodea canadensis.*

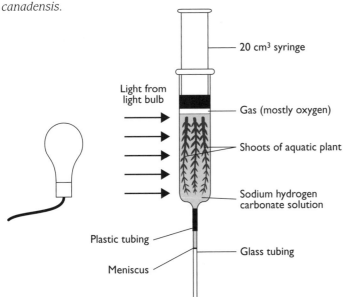

Figure 21 Apparatus for determining the rate of photosynthesis

When setting up the apparatus the following precautions must be taken:

- The water should be aerated so that it is saturated with oxygen. This ensures that any oxygen produced by the plant does not dissolve but comes out of solution to form bubbles that can be collected.
- The water contains some sodium hydrogencarbonate to provide a supply of carbon dioxide.
- When investigating one factor the others must be kept constant.

Bubbles of gas collect at the top of the syringe and the pressure forces the solution down the glass tube. The rate of photosynthesis is calculated by dividing the distance travelled by the meniscus by the time taken.

The light intensity is changed by placing the lamp at different distances from the plant or by putting grey filters between the lamp and the plant. The light intensity is measured with a light meter or is estimated by calculating $1/d^2$, where d equals the distance between plant and lamp. Why is the intensity $1/d^2$? Imagine putting a piece of card in front of a light source so the light covers an area $1\,cm^2$. When you move the card away the light is 'spread out' over a larger area. If you double the distance from the source, the light from the square is now 'spread out' over an area of $4\,cm^2$ (= 2^2) squares. At twice the original distance, the intensity of the light passing through a single square of $1\,cm^2$ is one-quarter of the original. At three times the original distance, the intensity of the light passing through a single square is one-ninth of the original. This is the inverse square law. The light intensity is proportional to $1/d^2$ (the inverse square of the distance).

The carbon dioxide concentration is changed by putting the plants into different concentrations of sodium hydrogencarbonate solution.

It is difficult to control the temperature with this apparatus or to use it to investigate the effect of different temperatures.

Example 21 (quantitative): leaf discs and photosynthesis

Discs of leaf tissue can be cut from seedlings of radish, *Raphanus sativus*. They are placed inside a syringe full of sodium hydrogencarbonate solution. The air is removed from inside the mesophyll of the discs by covering the nozzle of the syringe and pulling the plunger. The air is replaced with sodium hydrogencarbonate solution so the discs sink when placed in a dish containing the same solution. Four discs are placed under a lamp and the time taken for each disc to fill with gas and rise to the surface is recorded. This is repeated using fresh discs at different light intensities. Some discs are left in the dark. A student's results are given in Table 25 and Figure 22.

skills guidance

Table 25 The effect of light intensity on the rate of photosynthesis of leaf discs of Raphanus sativus

Distance from light, d/cm	Temp/ °C	Light intensity, $1000/d^2/$ cm^{-2}	Time taken for leaf discs to rise/s					Rate of photosynthesis/ s^{-1}
			1	2	3	4	Mean	
5	29	40.0	162	167	187	198	179	5.6
10	28	10.0	71	206	220	230	219	4.6
15	26	4.4	236	251	277	286	263	3.8
20	25	2.5	272	432	476	504	471	2.1
25	23	1.6	110	572	590	598	587	1.7

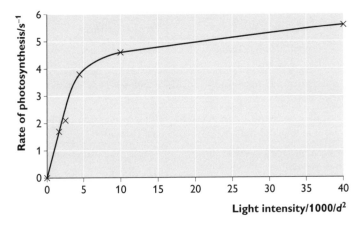

Figure 22 The effect of light intensity on the rate of photosynthesis of leaf discs of R. sativus

Figure 22 shows the expected trend that as light intensity increases so does the rate of photosynthesis. The student decided to include the origin since the leaf discs kept in the dark did not rise to the surface. Photosynthesis will not take place in the dark so the student was confident that the origin should be included. Note that the values of both the independent and dependent variables have been processed.

However, there are problems with controlling variables. In Examples 20 and 21 there are the following difficulties:
- There may be light from sources other than the bench lamps, so the values for light intensity are underestimates.
- It is difficult to know the concentration of sodium hydrogencarbonate in the water
- Oxygen released by photosynthesis dissolves in water, so early results may be underestimates unless the water has been aerated fully.
- Heat from lamps may increase the temperature of the water while taking readings.

Respiration

Example 22 (qualitative): vital stains and respiration

Neutral red and methylene blue are known as vital stains because they enter cells without killing them. Neutral red is a pH indicator. In alkaline conditions it is yellow; in neutral and acidic conditions it is red. Methylene blue is decolourised when reduced and turns blue when oxidised.

Table 26 shows some observations on the colours of these two vital stains when used with suspensions of yeast cells.

Table 26 Observations on suspensions of yeast made with two vital stains —
neutral red and methylene blue

	Treatment	Observations
1	Yeast cells suspended in alkaline solution and mixed with neutral red	Yeast suspension turns red
2	Yeast and neutral red suspension filtered	Residue is red; filtrate is colourless
3	Yeast and neutral red suspension boiled	Yeast suspension changes colour from red to yellow
4	Yeast cells suspended in water mixed with methylene blue	Yeast suspension turns blue
5	Glucose mixed with yeast and methylene blue suspension and kept warm	Blue colour gradually disappears
6	Decolourised suspension shaken thoroughly	Blue colour reappears
7	Boiled yeast suspension mixed with methylene blue	Yeast suspension turns blue
8	Glucose mixed with boiled yeast and methylene blue solution and kept warm	Yeast suspension stays blue and does not decolourise as in treatment 5

Try this yourself

21 Explain the results shown in Table 26.

Example 23 (quantitative): determining rates of respiration

Simple respirometers (Figure 23) can be used to investigate respiration rate.

skills guidance

Type A

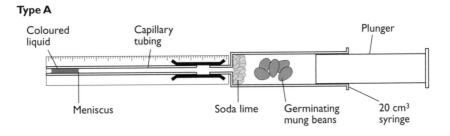

Type B

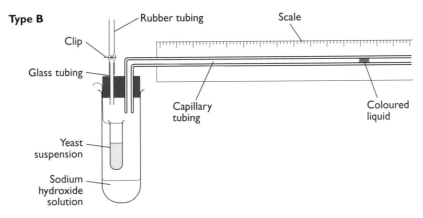

Figure 23 Two simple respirometers

Respirometers of types **A** and **B** should be left for at least 5 minutes to adjust. The soda lime and sodium hydroxide solution absorb carbon dioxide. When beans respire they use oxygen and produce carbon dioxide. In the type **A** respirometer, the carbon dioxide produced is absorbed by the soda lime so the volume of air in the syringe decreases. This decreases the pressure so that it is less than the atmospheric pressure, which causes the droplet of coloured liquid to move. The speed of movement of the droplet is a measure of the rate of respiration.

Controls may be set up by placing the carbon dioxide absorbent in the respirometers in the absence of the organisms or tissue. Any movement of the droplet in the control tube is due to a change in pressure, which may be the result of a change in temperature. The results from the respirometers can be adjusted to take this into account.

The respirometers can be kept in different places to investigate the effect of temperature on the rate of respiration — for example, in a refrigerator, at room temperature and in incubators set at different temperatures.

Readings should be taken at intervals to find out when the rate has become constant.

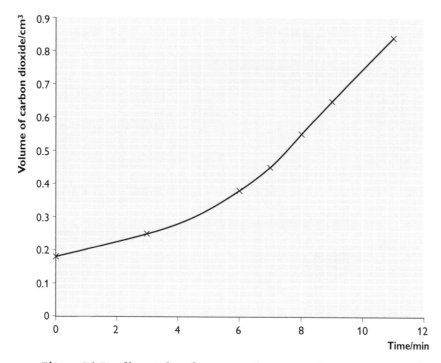

Figure 24 Readings taken from a respirometer of type B to determine the rate of respiration of yeast

Try this yourself

22 Use Figure 24 to calculate the rate of respiration of yeast.

Respirometers with carbon dioxide absorbent (Figure 23, Type A) are used to measure the rate of oxygen uptake in aerobic respiration. If the absorbent is omitted, then the production of carbon dioxide can be measured. If there is no movement in the droplet then the rate of oxygen uptake is the same as the rate of carbon dioxide produced. If the droplet moves to the right then less carbon dioxide is produced than oxygen absorbed. If the droplet moves to the left, then more carbon dioxide is being produced than oxygen absorbed. If respiration is anaerobic then movement of the droplet is to the left and there is no movement to the right, even with the carbon dioxide absorbent present.

Table 27 shows a student's results when using a type A respirometer (Figure 23). In both cases the meniscus moved towards the respiring seeds.

Table 27 Measurements from a type A respirometer of with and without carbon dioxide absorbent

Time/s	Distance moved by meniscus/mm	
	With soda lime	Without soda lime
0	0.0	0.0
30	12.5	4.5
60	21.5	3.5
90	33.0	6.0
120	42.5	9.5
150	54.0	16.0
180	64.0	22.0
210	71.5	24.5
240	79.5	25.5
270	88.5	25.5

Try this yourself

23 Use the results in Table 27 to calculate the rates of oxygen uptake and carbon dioxide production.

24 Explain why the rates of oxygen uptake and carbon dioxide production are not the same.

You can expect to carry out quantitative experiments on respiration using apparatus similar to those shown in Figure 23 and in Figure 3 in the Practical Tasks section (p. 110), to collect, analyse and interpret results and then to evaluate the procedure and the results.

As part of the processing you can expect to calculate volume changes by using the formula for the volume of a cylinder, V:

$$V = \pi r^2 d$$

where r = the radius of the capillary tubing; d = the distance travelled by the bubble or meniscus.

When investigating the effect of temperature on the rate of respiration you can calculate the temperature coefficient (Q_{10}) (see p. 112 in the A2 quantitative task).

Example 24 (quantitative): respiration and respiratory substrates

The effect of different substrates on respiration can be investigated.

Yeast is mixed with different solutions and drawn up into syringes. These have nozzles fixed to the end and are left to drip as shown in Figure 25. If yeast is incubated with different respiratory substrates the presence of droplets can be used to assess whether the yeast is able to respire the substrate. The results obtained are shown in Table 28.

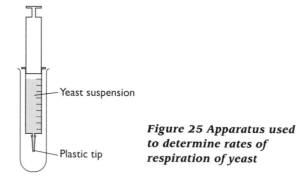

Yeast suspension

Plastic tip

Figure 25 Apparatus used to determine rates of respiration of yeast

Table 28 The effect of different sugars on the respiratory rate of yeast

| Tube | Respiratory substrates | Number of drops per minute | | | | Volume of gas in syringe barrel after 5 minutes/cm³ |
		1	2	3	4	
A	Glucose	17	15	16	20	5.65
B	Fructose	14	15	17	18	4.70
C	Galactose	2	2	1	0	0.60
D	Maltose	14	15	17	22	4.40
E	Lactose	2	2	0	0	0.70
F	Sucrose	12	13	17	20	5.65
G	Water	2	1	0	0	0.65

Try this yourself

25 How would you present the data in Table 28?

26 What can you conclude from these results?

Example 25 (evaluative): effect of temperature on respiration

Triphenyltetrazolium chloride (TTC) is an artificial hydrogen acceptor that turns pink when reduced. It is a redox dye. Put 10 cm³ of a yeast suspension prepared in glucose into a test tube and keep in a water bath at 65°C. Put 1 cm³ TTC solution into another test tube and put this into the water bath. After 3 minutes add the yeast suspension to the TTC solution, mix and return to the water bath. Start a stop watch and record the time it takes for a pink colour to appear. Repeat the procedure at five lower temperatures.

Some student results are shown in Table 29.

skills guidance

Table 29 Determining rate of respiration in yeast using TTC

| Temperature/°C | Time taken for pink colour to appear/s | | | |
	Student A	Student B	Student C	Student D
25	455	205	274.12	288
35	190	120	118.13	148
45	112	57	59.50	48
55	64	36	17.37	32
65	93	38	20.00	269

Try this yourself

27 Comment on the reliability of the results in Table 29.

28 How would you process and present the data in Table 29?

29 What can you conclude from these results?

Example 26 (evaluative): photosynthesis and respiration

This example shows two types of calculation you may be asked to do in an evaluative task that uses results from photosynthesis and respiration.

It is only possible to measure the **apparent rate of photosynthesis**. Some of the oxygen produced in photosynthesis is used in respiration within the plant and is not released. Rates of photosynthesis in aerial plants are determined by using gas analysers that measure the uptake and/or release of carbon dioxide. This presents the same problem — the quantity of carbon dioxide absorbed by a plant is less than the total quantity used in photosynthesis because chloroplasts fix carbon dioxide released by mitochondria in respiration.

To find the **true rate of photosynthesis** it is necessary to first find out the rate of respiration for the conditions used. Table 30 shows how this was carried out in an investigation in which the uptake and release of carbon dioxide were measured at different temperatures.

Table 30 Determining the true rate of photosynthesis by adding the rate of respiration to the apparent rate of photosynthesis

Temperature/°C	5	10	15	20	25	30
Apparent rate of photosynthesis: net uptake of CO_2 in **bright light**/ mgg (dry mass) $^{-1}h^{-1}$	1.3	2.4	3.0	3.3	3.0	2.2
Rate of respiration: release of CO_2 in **dark**/mgg (dry mass) $^{-1}h^{-1}$	0.4	0.7	1.0	1.4	1.9	2.8
True rate of photosynthesis: total use of CO_2 in photosynthesis/ mgg (dry mass) $^{-1}h^{-1}$	1.7	3.1	4.0	4.7	4.9	5.0

It is possible to calculate the temperature coefficient Q_{10} from the figures in Table 30. The temperature coefficient is the relative change in the rate of a process with an increase in temperature. Q_{10} is the relative change for a temperature increase of 10°C.

$$Q_{10} = \frac{\text{rate of process at temperature } (t + 10°C)}{\text{rate of process at temperature } t}$$

If you calculate these changes for respiration using information in Table 30 for 5°C to 15°C, 10°C to 20°C, and so forth you will discover that the values are positive and near to 2, up to 30°C. This is the value expected for Q_{10} for enzyme-catalysed reactions. On reaching the optimum temperature the Q_{10} is nearer 1. After the optimum temperature, Q_{10} is negative as enzymes are denatured. You do not see the same relationship with photosynthesis because it is influenced by the availability of light in the light-dependent stage as well as by temperature in the light-independent stage.

Excretion

Example 27 (qualitative): problem solving — excretion

Part 1

You are provided with three solutions, **A**, **B** and **C**, which have been prepared to resemble different body fluids — glomerular filtrate, blood plasma and urine. Use the reagents and test strips provided to identify the samples of the body fluids. Explain your identifications in terms of the contents of the three fluids. The following materials are provided:

- Clinistix®, which detects glucose
- Albustix®, which detects protein
- Urease solution and red litmus paper — urease hydrolyses urea to ammonia and carbon dioxide; ammonia vaporises and is detected by red litmus paper

Table 31 Results of tests on three samples of body fluids, A, B and C

Samples of body fluids	Observations		
	Clinistix®	Albustix®	Urease and red litmus paper
A	No change	No change	Red → blue
B	Blue → purple	No change	Red → blue
C	Blue → purple	Yellow → green	Red → blue

Try this yourself

30 Use the information in Table 31 to identify the body fluids, A, B and C.

Part 2

Urine tests may be carried out as part of a medical investigation. Read the following information and then use the materials provided to test samples **D** to **J** which have been prepared to resemble urine samples.

Urine tests and the diagnosis of some diseases

During aerobic respiration, glucose is oxidised to pyruvate, which is converted to acetyl coenzyme A. This condenses with oxaloacetate so that a two-carbon fragment (the acetyl group) enters the Krebs cycle. Stored triglycerides are hydrolysed to glycerol and fatty acids. Fatty acids are oxidised to acetyl coenzyme A, in the process known as β-oxidation, so that two-carbon fragments derived from fat may enter the Krebs cycle.

In **diabetes mellitus** and in people who are dieting, fasting or starving, oxaloacetate is not available because it is used to form glucose. This formation of 'new' glucose is called gluconeogenesis. Under these conditions, the acetyl group from acetyl coenzyme A is converted to substances called ketones that may appear in the urine.

In **chronic nephritis**, the glomeruli show progressive destruction with the result that they become less efficient at filtering the blood. Nephritis is often associated with bacterial infections of the kidneys. Bacteria may also infect the urinary tract of people with kidneys that function properly. The presence of nitrite ions in urine indicates that bacteria may be present in the urinary system in large numbers.

Bile salts decrease the surface tension of water. If powdered sulfur is put on the surface of water it floats because of surface tension. It sinks rapidly if there are any emulsifiers in the water. In **obstructive jaundice**, the bile duct is blocked and the constituents of bile enter the blood, are filtered and excreted in the urine.

If little or no antidiuretic hormone (ADH) is secreted large volumes of very dilute urine are produced in the condition known as **diabetes insipidus**.

Procedure

Use the powdered sulfur, test strips and universal indicator solution to test the urine samples. Compare the colours of the samples. In each case, suggest a likely diagnosis.

Table 32 Results of tests on six urine samples, D to J

Test/ observation	Urine samples					
	D	E	F	G	H	J
Diastix®	✓	✗	✗	✗	✗	✗
Albustix®	✗	✓	✗	✗	✗	✗
Ketostix®	✓	✗	✗	✗	✗	✓
Nitrite test strip	✗	✓	✗	✗	✗	✗
pH	6.0	9.0	6.5	6.5	6.0	4.0

Test/	Urine samples					
observation	D	E	F	G	H	J
Powdered sulfur	Floats	Floats	Floats	Floats	Sinks	Floats
Colour	Pale yellow	Pale yellow	Pale yellow	Very pale yellow	Pale yellow	Pale yellow

Try this yourself

31 Use the results in Table 32 to match the urine samples, **D** to **J**, with the conditions described above.

32 Explain the presence of protein and glucose in the urine.

Biological drawings

There are four topics from the specification for F214 and F215 that could provide material for making drawings:
- histology of the liver
- histology of the kidney
- histology of the pancreas
- sections of anthers to show meiosis

Example 28 (qualitative): high-power drawing of the pancreas

Figure 26 shows an annotated drawing made from a slide of the pancreas.

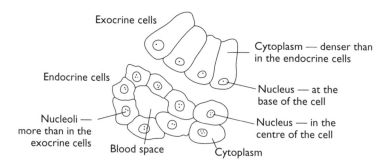

Figure 26 Annotated drawing of exocrine and endocrine cells from the pancreas

Note that in the exocrine tissue:
- the cells are arranged closely together; there are no gaps between them
- the nucleus is at the base of each cell
- the cytoplasm is dense

In the endocrine tissue:
- the cells are arranged loosely; there are gaps between the cells which are blood spaces
- the nucleus is at the centre of each cell
- the cytoplasm is less dense than in the exocrine tissue

Exocrine and endocrine cells are protein-synthesising cells. There is information about the structure and function of such cells in all four units:
- function of organelles in protein synthesis (F211)
- structure of proteins (F212)
- role of cells in the islets of Langerhans (F214)
- transcription and its control (F215)
- translation (F215)

There are a number of synoptic questions that could be asked about the structure and function of these cells.

You are not likely to be asked to make a preparation to show meiosis as you could be for mitosis. However, you could be provided with prepared slides or photographs and be expected to identify the stages and draw the chromosomes. There are suitable photographs at: **http://images.iasprr.org/lily/male.shtml**

Example 29 (qualitative): products of meiosis

It is possible to see the results of meiosis by studying the products of meiosis in the fungus *Sordaria fimicola*. In this species, as in many fungi, meiosis occurs after fertilisation to produce spores. Immediately after meiosis a mitotic division occurs. The spores form in a tube-like structure known as an ascus and are lined up in the order in which they form (see Figure 27). There is a gene in *S. fimicola* that controls the colour of the ascospores; the allele **B** gives rise to black spores and the allele **b** gives rise to white spores. The spores are aligned in a 4+4 arrangement if there has been no crossing over between the **B/b** locus and the centromere. Any other arrangement (e.g. 2+2+2+2) shows that crossing over must have occurred.

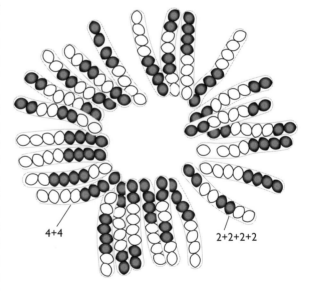

4+4

2+2+2+2

*Figure 27 Drawing made from a photomicrograph of asci of **S. fimicola***

> **Try this yourself**
>
> **33** Count the number of asci in which there is a 4+4 arrangement of black and white ascospores.
>
> **34** Count the number of asci in which there is any other arrangement of ascospores.
>
> **35** From your results, calculate the percentage frequency of crossing over during meiosis in *S. fimicola*.

You can read more about this and carry out another exercise on crossing over in *S. fimicola* at: **http://www.phschool.com/science/biology_place/labbench/ lab3/design2.html**

Application of statistics

Example 30 (quantitative): discontinuous variation

Discontinuous variation is shown by investigating the inheritance of eye colour in the fruit fly, *Drosophila melanogaster*. The locus for one of the genes for eye colour is on the X chromosome. The alleles of the gene are **R** for red and **r** for white. In this example, the flies in the parental generation were pure-bred red-eyed males and pure-bred white-eyed females. Students are provided with photographs of the results of two crosses. Table 33 shows the results of counting the different types of flies in the photographs.

Table 33 Counts of red-eyed and white-eyed fruit flies from photographs of F₁ and F₂ offspring

Phenotype	F₁ Tally	Total	F₂ Tally	Total
Red-eyed males		0	ⵏ ⵏ ⵏ II	17
White-eyed males	ⵏ ⵏ ⵏ ⵏ ⵏ IIII	29	ⵏ ⵏ ⵏ III	18
Red-eyed females	ⵏ ⵏ ⵏ ⵏ ⵏ ⵏ ⵏ IIII	39	ⵏ ⵏ ⵏ ⵏ ⵏ	25
White-eyed females		0	ⵏ ⵏ ⵏ ⵏ ⵏ III	28
Totals		68		88

> **Try this yourself**
>
> **36** State a null hypothesis for each cross and use the chi-squared (χ^2) test to see if the differences between the observed and expected results are significant.

Biotechnology

Example 31 (qualitative): immobilised enzymes and cells

One method of immobilising cells and enzymes is to encapsulate them in small beads of calcium alginate, which is a very viscous substance. Sucrase, amylase, lactase and urease can be immobilised and tested to find how efficient they are compared with non-immobilised enzymes. They may be slower in action as substrates have to diffuse through the gel to reach the active sites of the enzymes. However, they can withstand higher temperatures and a wider pH range; they can be used in columns; they can be filtered easily from the product and, therefore, not lost from the production process so they can be reused.

There are a number of readily available enzymes that can be immobilised and used to investigate the effects of immobilisation, such as contamination of the product and its rate of production.

Yeast can also be immobilised and its enzymes used — for example, sucrase and catalase using sucrose and hydrogen peroxide as substrates. The advantage of using cells is that they can carry out multistep processes involving many enzymes. The disadvantage is that they use some of the substrate for their own requirements, rather than converting it directly into product. The A2 quantitative task on p. 105 is an example of an investigation using an immobilised enzyme.

Behaviour

Innate behaviour in invertebrates can be investigated using simple apparatus. Two forms of innate behaviour are kinesis and taxis. These are good topics for qualitative tasks. Examples 32 and 33 show how they can be investigated using woodlice and blowfly larvae (maggots).

Example 32 (quantitative): kinesis

Figure 28 shows a choice chamber. This can be used to give two halves that have different conditions, such as light versus dark, damp versus dry and combinations of conditions such as light and dry versus dark and damp. Woodlice, usually of the genera *Porcellio* and *Oniscus*, are released into the choice chamber. Results can be collected in a number of ways. The total number of woodlice in each compartment or the number that are moving in each compartment can be counted at timed intervals. You may be expected to choose a way to record the behaviour of the animals and then devise a table to record your observations.

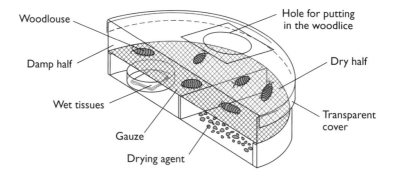

Figure 28 A choice chamber

Figure 29 shows the types of behaviour shown by woodlice and other invertebrate animals.

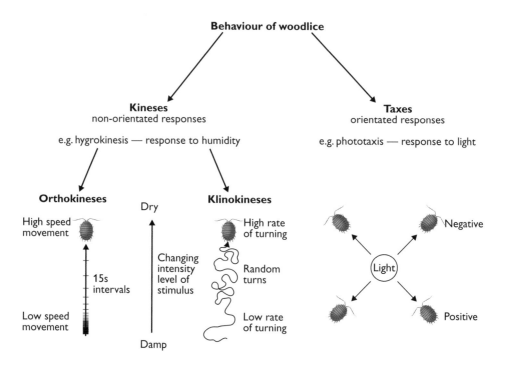

Figure 29 Types of behaviour shown by woodlice

skills guidance

Table 34 shows some data from choice-chamber investigations

	Investigation 1		Investigation 2	
	Number of woodlice in each half of choice chamber		Number of woodlice moving in each half of the choice chamber	
Time/min	Light	Dark	Dry	Humid
0	10	0	5	5
1	9	1	8	2
2	6	4	7	1
3	5	5	5	1
4	3	7	3	2
5	3	7	1	2
6	3	7	1	0
7	2	8	1	0
8	1	9	0	0
9	1	9	0	0
10	1	9	0	0

Try this yourself

37 How would you present and analyse the data in Table 34?

38 State the type of behaviour shown by the woodlice, justifying your answer by reference to the data.

Example 33 (quantitative): taxis

Blowflies lay their eggs in carcasses of rotting meat. The eggs hatch and the larvae (maggots) feed on the meat. The maggots then develop through several stages before pupating and then changing into adult flies. Development from egg to adult occurs in 16 to 35 days, depending on temperature and other environmental conditions. The larvae are very active, particularly towards the time when they pupate. Later they emerge from the carcass, and crawl away to pupate, usually in the soil or a safe crevice. Figure 30 shows the apparatus that can be used to investigate the behaviour of maggots.

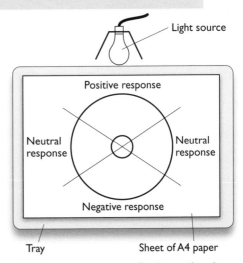

Figure 30 Apparatus for investigating responses of maggots to light

These are the instructions to follow when using this apparatus:

(1) The desk lamp is about 100 mm from one side of the tray, shining at a shallow angle along the tray. The marked A4 paper is placed in the tray.

(2) Place one active maggot at the exact centre of the small circle and place a transparent cover over the tray. Mark the position of the centre on the cover.

(3) Start a stopwatch as soon as the head of the maggot crosses the inner circle. Mark this position with a small 'a'. Every 10 s, mark the maggot's position again, using the same symbol. If the maggot moves quickly, then 5 s or 2 s may be a better interval.

(4) Record across which sector (positive, negative, or neutral) the maggot leaves the outer circle. Continue marking its position at regular intervals until it reaches the edge of the marked paper.

(5) Repeat steps 2–4 until you have recordings from 10 maggots: use a new symbol ('b', 'c', 'd', etc.) for each one. Take care to place the cover in the same position for each trial.

(6) Place the cover on a large sheet of paper and mark the relative position of the lamp. For each maggot, make a table of the distance from the lamp to each position, and the speed of movement (in $mm\,s^{-1}$) from that position to the next.

Table 35 Summary of results from investigating behaviour of maggots using the apparatus shown in Figure 30

Sector	Number of maggots	Mean speed at 2 s intervals/$mm\,s^{-1}$				
		2	4	6	8	10
Positive	1	1.75	0.75	1.0	0.0	0.0
Negative	15	12.4	9.5	8.3	5.5	3.5
Neutral	4	1.56	0.59	0.35	0.25	0.0

Try this yourself

39 How would you analyse the data in Table 35?

40 Comment on the behaviour shown by the maggots, justifying your answer by reference to the data.

Fieldwork

Various aspects of ecology are covered in F212 and F215 and ecological investigations could be included in F213 and/or F216. Investigations may be completed at field centres or during fieldwork at suitable sites near your school or college. Here are some methods that are used to investigate distribution and abundance of organisms, using the rocky shore, sand dunes and salt marshes as examples.

A **timed search** can be used to find as many different species as possible so that a species list can be compiled. Walk across a delineated area searching for different species. You might be concentrating on a single group of organisms such as seaweeds or molluscs. The search may be carried out in a specific area and a dichotomous key used to identify organisms. This gives an indication of the **species richness** of the area. Counting the number of individuals of each species gives an indication of the **species evenness** and is used to calculate **Simpson's index of diversity**.

Example 34 (qualitative and quantitative): random sampling

The problem with a timed search is that the sampling may not be random. You may be sampling parts of the area that are species rich (or not!). To overcome any bias, random sampling is used. Tape measures are used to make a grid, random numbers are used to give coordinates and quadrats are placed on these coordinates. The abundance of different species in a qualitative task would be done by using an abundance scale, such as ACFOR. In a quantitative task, the abundance may be recorded by:

- counting individuals (e.g. limpets, periwinkles, and top shells as in Table 36)
- estimating percentage cover (e.g. seaweeds and barnacles which are difficult to count individually)

These data can be used to assess species evenness. They might also be analysed to see if there is any association between species. This method of sampling is usually done in a uniform ecosystem, such as grassland or woodland. The three ecosystems mentioned above (rocky shore, sand dunes, salt marsh) all show changes in space and may also show changes in time.

Table 36 Data collected on molluscs on a rocky shore

Species of mollusc	Near to a stream	Away from a stream
Patella vulgata	4	8
Littorina littorea	41	25
Littorina saxatilis	32	45
Littorina obtusata	0	4
Gibbula umbilicalis	0	36
Osilinus lineatus	54	38
Total	131	156

Try this yourself

41 The data in Table 36 were collected near to, and away from, a stream that rises above the high water mark and runs across the rocky shore into the sea. Use the data to calculate Simpson's index of diversity for the two areas.

42 Comment on the results.

Example 35 (quantitive): association between species

The association between two species can be investigated by using a contingency table.

Students investigated the influence of a biotic factor on the distribution of flat periwinkles, *Littorina mariae*. At low tide, these molluscs settle on the fronds of seaweeds, most frequently on serrated wrack, *Fucus serratus*. Students noted where the colour morphs, yellow and brown, settled on the serrated wrack — either on the stipe and holdfast at the base of the seaweed or on the tips of the fronds.

The data collected were organised into a contingency table as shown in Table 37. In the table, the expected numbers, assuming no association between the colour morphs and position on the serrated wrack, are given in brackets.

Table 37 Contingency table for data on distribution of flat periwinkles

Colour morphs	Settling position on serrated wrack		Row totals
	Stipes and holdfasts	**Tips of fronds**	
Yellow	82 (107)	142 (117)	224
Brown	179 (154)	145 (170)	324
Column totals	261	287	**Grand total** 548

Does this distribution occur by chance or is there an association between colour morphs and the places where they settle on serrated wrack? The data collected are categoric. We can use the data to calculate the expected numbers assuming no association — for example, for the yellow morphs on the tips of the fronds:

$$\frac{\text{column total} \times \text{row total}}{\text{grand total}} = \frac{287 \times 224}{548} = 117$$

Try this yourself

43 Use the chi-squared test to find out whether there is a significant difference between the observed and expected results in Table 37. (Degrees of freedom = (number of columns – 1) × (number of rows – 1)).

44 Comment on your answer.

Example 36 (quantitative): line transects and abiotic factors

A long rope marked at regular intervals (e.g. every 0.5 m) is placed across the shore and the organisms found at regular intervals recorded. The results are converted into a drawing that shows the **distribution** of organisms. Line transects are used to show how communities change along a gradient, which could be a slope or a change in an abiotic feature. They are a good way to show the changes qualitatively. You can see an example of a line transect across a wetland ecosystem at:

www.countrysideinfo.co.uk/wetland_survey/transec1.htm

Data on abiotic factors can also be collected at sampling points along transects. Factors may include:

- light intensity
- temperature
- humidity of air (measured using a hygrometer)
- wind speed (measured using an anemometer)
- oxygen concentration, pH and salinity of water
- pH and water content of soil

Abiotic factors must be recorded consistently — for example at the same height or depth. The results for the abiotic factors are written above or below the line transect.

You may be asked to collect such data and construct a line transect as a qualitative task. You could do this somewhere where primary succession exists, such as wetland, salt marsh or sand dunes, or where there is an abrupt change in habitat, for example, between a field and adjoining woodland.

Example 37 (quantitative): continuous variation

Table 38 (a) shows the heights of dog whelks collected on a sheltered shore and an exposed shore. Figure 31 shows how the shell height is measured. To be displayed as a histogram the data have first to be divided into classes and the number in each class scored. This is shown in Table 38 (b).

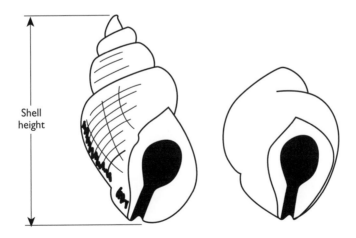

Shell height

Figure 31 Dog whelks from a sheltered shore and an exposed shore

Table 38 (a) Raw and processed data for dog whelks

	Height of dog whelk/mm									
	Sheltered shore					Exposed shore				
	23	28	29	32	35	13	16	18	21	25
	23	29	29	32	36	14	17	18	21	25
	26	29	29	32	37	14	17	19	21	25
	26	29	30	33	37	14	17	19	22	25
	26	29	30	33	37	15	17	19	23	25
	27	29	30	34	38	15	17	20	23	26
	28	29	30	34	38	15	17	20	23	27
	28	29	31	35	39	15	17	20	24	27
	28	29	32	35	39	16	18	20	24	28
	28	29	32	35	39	16	18	20	24	28
Total number	50					50				
Range	23–39 = 16					13–28 = 15				
Mean	31.3					14.8				
SD	4.1					4.2				

Table 38 (b) Dog whelk: organised data

Sheltered shore			Exposed shore		
Height / mm	Tally	Total	Height / mm	Tally	Total
23–24	II	2	13–14	IIII	4
25–26	III	3	15–16	₥ II	7
27–28	₥ I	6	17–18	₥ ₥ I	11
29–30	₥ ₥ ₥ I	16	19–20	₥ III	8
31–32	₥ I	6	21–22	IIII	4
33–34	₥	5	23–24	₥ I	6
35–36	IIII	4	25–26	₥ I	6
37–38	₥	5	27–28	IIII	4
39–40	III	3	29–30		0
Total		50			50

This information is plotted on the histogram in Figure 32. The most frequent class is the mode. A statistical test can be used to show that the difference between the two populations is significant and not due to chance alone.

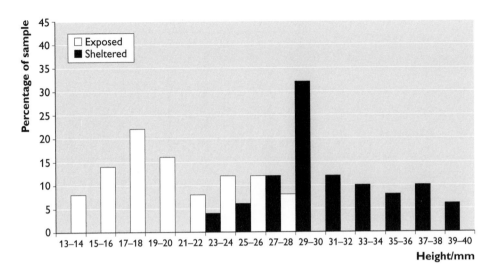

Figure 32 Histogram to show variation in shell height in dog whelks from sheltered and exposed shores

The relationship between a feature that shows continuous variation and an abiotic factor can be shown by plotting a scattergraph with the abiotic factor on the *x*-axis.

A correlation may be positive or negative. The strength of the correlation can be assessed by calculating the correlation coefficient. A coefficient of +1 indicates there is a perfect positive correlation (all the points are on a straight line); a coefficient of –1 indicates that there is a perfect negative correlation. A coefficient of 0 means the points are scattered at random on the graph.

You can often show an association between two species but cannot prove cause and effect. There may often be another factor or factors that determine the association — there are so many factors that it is impossible to measure all of them. The results obtained are 'snapshots' of the ecosystem at a particular time. The distribution and abundance of species changes during the day, between seasons and from year to year.

Example 38 (quantitative): belt transects

A **continuous belt transect** is carried out by placing frame quadrats or point quadrats all the way along the transect and recording the organisms. This is extremely time consuming. Placing quadrats at intervals gives an **interrupted belt transect**. Data from a belt transect can be used to construct a kite diagram. Table 39 shows some data collected by students on a rocky shore. They began at low water and placed quadrats at 20 m intervals.

Distance/ m	Seaweeds				Molluscs				
	Spiral wrack	Bladderwrack	Serrated wrack	Kelp	Rough periwinkle	Flat periwinkle	Edible periwinkle	Limpet	Dog whelk
200	0	0	0	0	0	0	0	0	0
180	0	0	0	0	5	0	0	0	0
160	0	0	0	0	5	0	0	0	0
140	5	5	0	0	0	0	0	0	0
120	4	1	0	0	0	0	0	0	0
100	0	5	5	0	0	0	0	0	0
80	0	5	5	0	5	4	0	5	0
60	0	5	5	0	4	0	1	4	0
40	0	5	5	0	0	0	4	4	0
20	0	5	4	0	0	0	4	5	1
0 (low water)	0	0	5	3	0	0	0	0	0

Table 39 Data collected by students from an interrupted belt transect

The data collected by the students is shown in the kite diagram in Figure 33. Figure 34 shows how to use data like those in Table 39 to draw a kite diagram.

Converting the ACFOR scale into a numerical scale:

ACFOR scale	Abundance scale
Species absent	0
Rare	1
Occasional	2
Frequent	3
Common	4
Abundant	5

Results for a mollusc species:

Distance along transect/m	ACFOR scale	Abundance scale
0	–	0
10	–	0
20	R	1
30	C	4
40	–	0
50	–	0
60	O	2
70	O	2

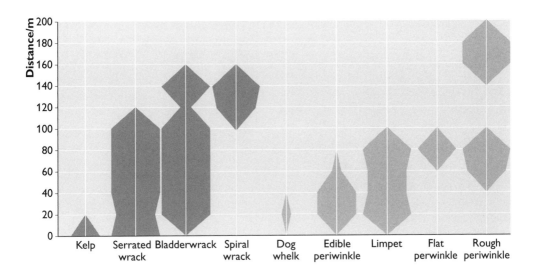

Figure 33 How to draw a kite diagram

Figure 34 The distribution and abundance of organisms along a transect is shown in a kite diagram

General points

You can expect to undertake fieldwork to collect data on populations, communities and abiotic factors. You are expected to know about sampling and to be able to explain your results and evaluate the procedures used and data collected. Note that rocky shores show zonation because the communities are determined by the height of the tides and do not change over time, as happens with successions on sand dunes and salt marshes. When evaluating ecological tasks expect to comment on the following problems:

- Identifying and counting different species may decrease the accuracy of the results.
- Sampling may be biased so the results are not representative.
- Too few samples may be taken so results may not be reliable.
- Distribution and abundance of organisms change over time — results may not be valid because they are a 'snapshot' taken at one place on one occasion and may not be reproducible.

Look at this website for information about marine habitats in the UK:
www.marlin.ac.uk

And these for information about sand dunes and the plants that grow on them:
www.sandsoftime.hope.ac.uk
www.habitas.org.uk/flora/habitats/sanddunes.htm

Other sites of interest are:
www.field-studies-council.org/resources/
www.countrysideinfo.co.uk/index.htm

Practical tasks

The questions in this section of the guide follow the sequence of tasks. When doing the tasks you will be given the following question papers:

- **Qualitative task** (this may be accompanied by further resource sheets)
- **Quantitative task part 1**: procedure and table to complete
- **Quantitative task part 2**: processing and analysis of data collected in Part 1
- **Evaluative task**
  - **Student experiment sheet (SES)**, which contains the procedure for a task that may either be identical to, or slightly different from, the quantitative task. The results in **Table 1** have been prepared by the examiner. You are expected to process, analyse and interpret these results. You are expected to evaluate the procedure.
  - **Question paper** on which you write your answers.

(For the evaluative tasks in this guide the details of the practical procedures have been omitted from the student experiment sheets.)

The question papers for the **Qualitative Task** and **Part 1** of the **Quantitative Task** include a list of the apparatus and materials that you should have in front of you when you start the task. To save space the apparatus lists have been omitted from the tasks in this guide. The apparatus that you use will probably be familiar to you; if it is not, then the procedure will give full instructions about how to use it. Lists of materials are included in the tasks in this guide.

The minimum for grade A is at least 80% of the maximum mark (in this case around 32 marks). Candidate A gains full marks for all the questions. Candidate B makes a lot of mistakes, often those that examiners encounter frequently. Candidate B's marks are given — if the overall mark is about 55% of the total (around 22 marks), then the candidate may gain a grade-E pass. Use these benchmarks when trying the questions yourself.

Candidates' answers are followed by typical examiner comments. These are preceded by the icon 𝒆 and indicate where credit is due. In Candidate B's weaker answers they also point out areas for improvement, specific problems and common errors, such as poor tabulation and presentation of results, lack of clarity, weak or non-existent development, irrelevance, misspellings, misinterpretation of the question and mistaken meanings of terms. However, some of Candidate B's answers are good — so read the comments carefully.

F213: Practical Skills in Biology 1

AS qualitative task: the effect of temperature on cell membranes

Background

Beetroot cells contain the red pigment betalain in their vacuoles. Betalain is water soluble and under normal circumstances does not diffuse out of vacuoles. Plant cell vacuoles are surrounded by a membrane, the tonoplast, which separates the contents of the vacuole from the cytoplasm. The tonoplast is similar in composition to the cell surface (plasma) membrane.

Introduction

For this practical task you will **investigate the effect of a range of temperatures on the release of betalain from fresh beetroot tissue**. You will make some standard solutions of betalain and use these to estimate the loss of this pigment from the beetroot tissue.

You are given full instructions for the practical procedure, which must be followed carefully. **Before you start any practical work, you are advised to read through the procedure carefully and study Table 1**.

It is your responsibility to organise your time efficiently and to work safely. You will be assessed on the accuracy of your results.

For this task you could be awarded 1 mark for demonstrating each of the following:
- safe working
- skilful practice

Safety

The experiment must be carried out safely.

Part 1

Materials
You are provided with the following:
- 10% betalain solution
- 1% betalain solution
- distilled water
- 50 discs cut from fresh beetroot tissue. Each disc is approximately 5 mm in thickness. These have been washed thoroughly in distilled water to remove any

trace of the pigment betalain. The discs are provided in five specimen tubes, 10 discs per tube, labelled with the temperatures that you will use (**30°C** to **70°C**)
- a beaker of **distilled water** at a temperature of 30°C (±2°C)

Procedure

Read steps **1** to **7** before you start your practical work.

1 Use the two 10cm³ syringes and the 1cm³ syringe provided to make up 10cm³ of each of the standard solutions in the labelled test tubes provided. Use the 10% betalain solution, the 1% betalain solution and water as shown in Table 1.

You will require these standard solutions for step **7**.

Table 1

Volume of 10% betalain solution/cm³	Volume of 1% betalain solution/cm³	Volume of distilled water/cm³	Concentration of standard solution of betalain/%
10	–	0	10.0
5	–	5	5.0
–	10	0	1.0
–	5	5	0.5
–	1	9	0.1
–	–	10	0.0

2 Pour off the water from the specimen tube labelled **30** into the beaker labelled **waste**. Rinse the discs in two changes of distilled water. Pour away the water you have used for rinsing into the beaker labelled **waste**.

3 Use a syringe to transfer 10cm³ distilled water at 30°C (±2°C) into the specimen tube so that the discs are covered. Start a stopwatch or stop clock and **leave it running** for the duration of the investigation.

4 Use the Bunsen burner to increase the temperature of the **distilled water** to 40°C (±2°C). Repeat step **3** with the discs in the specimen tube labelled **40**. Use a syringe to transfer 10cm³ distilled water at 40°C into the tube labelled 40. Immediately record the time from the stop clock or stopwatch.

5 Repeat step **4** with water at 50°C, 60°C and 70°C (±2°C in each case).
When the discs covered with water at 70°C have been immersed for **at least 10 minutes**, proceed with step **6**.

6 Pour off the water from each specimen tube **into the appropriate test tube** labelled 30, 40, 50, 60 or 70.

7 Estimate the concentration of betalain in each test tube by comparing the intensity of the colour against the standard solutions that you prepared in step **1**.

Construct and complete a table to record your observations and estimates.

(4 marks)

Part 2

1 Explain why it was necessary to wash the beetroot discs before immersing them in water at different temperatures (step **3**) *(2 marks)*

2 Suggest an explanation for the results you have recorded. *(2 marks)*

Total: 10 marks

Candidates' answers to the AS qualitative task: part 1

✐ The teacher awarded the marks for *safe working* and *skilful practice* to both candidates. This is because they wore lab coats while doing the practical work and kept tidy, well organised work spaces on the bench and did not spill any of the betalain solutions. They worked carefully through the procedure following the instructions about the timing and the use of the standard solutions to estimate the concentrations of betalain. Both were careful to use the labelling pen provided to label the standard solutions with the concentrations given in Table 1.

Note that the marking points that your teachers will use when assessing skilful practice and safety are not the same for each task. The examiners write marking points that are appropriate for each task. However, you should always pay attention to aspects of safe practice in the laboratory and take care to follow the practical instructions in each task very carefully. If you do this you are likely to gain the marks available.

Candidate A

Temperature/°C	Colour of water around the discs	Estimate of betalain concentration/%
30	Colourless	< 0.1
40	Very light pink	< 0.1
50	Darker pink than 40°C	Between 0.1 and 0.5
60	Pink	0.5
70	Dark pink	>0.5, <5.0

Candidate B

Temperature	Observation	Concentration	Change to the beetroot cells
30°C	Clear water	0	Cells are not leaky
40°C	Clear with hint of pink	0.1	Ditto
50°C	Slightly pink	0.1	Cells leak some pigment
60°C	Pink	0.5	Cells are partially leaky
70°C	Most pink	5.0	Cells are very leaky

ℯ This task is based on a learning outcome from Unit F211. You can read about this on p.21 of the Unit Guide for F211.

Candidate B has put the independent variable (temperature) in the left-hand column, but has put the units (°C) beside each number instead of in the heading. Units should not appear after numbers in tables, they should be given only in the heading either after a solidus or in brackets. Candidate B has not given an outer border to the table and has not given informative headings to the second and third columns. The fourth column should not be included because it gives *conclusions* not observations or estimates of the betalain concentration of the bathing water. However, Candidate B's results follow the pattern expected in that the colour at 70°C is a darker pink than that at 40°C. The estimates do not include at least one range as expected by the examiners. Do not put 'ditto' or 'as above' in a table. Each row must have an observation, in this case about the colour observed. Candidate A has estimated the concentration at 60°C as an exact concentration from the colour standards. All the others are given as ranges. Candidate A has decided that the colours at 50°C and 70°C are between the colour standards. This is quite likely to happen in a task of this type. Note that it is useful to use the symbols < (less than) and > (greater than) when recording ranges in a table. Candidate B gains 1 mark for obtaining the correct trend in the colours of the bathing solutions.

Part 2

Candidate A

1 The discs were washed to remove all the betalain pigment. As the discs are cut the cells are broken open and the betalain stays on the surface of the discs. If they were not washed then it would be impossible to tell how much has leaked out of the discs during the time they are in the hot water. If the pigment was not washed away all the concentrations would be overestimated.

Candidate B

1 Washing the discs removes any contaminants so they do not ruin the experimental results.

ℯ Candidate A's detailed answer explains the effect that not washing the discs would have on the results. Candidate B does not show any understanding of the reason for this part of the procedure. 'Removes contaminants' may be a suitable answer in other practical contexts, but not here.

Candidate A

2 In water at 50°C to 70°C betalain has diffused out of the vacuoles, through the tonoplasts, cytoplasm and cell membranes. At and below 40°C the membranes are intact, but above 50°C the membranes break down so the betalain escapes.

This is because proteins in the membrane are denatured and phospholipids become more fluid.

Candidate B

2 At high temperatures, proteins are denatured so betalain leaves the discs and enters the water. The proteins in the membranes break down so betalain is not kept in the vacuoles. This happens at the highest temperatures.

> Candidate A uses the appropriate terminology from F211 and has given a concise explanation. Candidate B realises that at the high temperatures in the range tested, proteins are denatured. However, these comments should be linked to one or more temperatures: 'At high temperatures' is not specific enough. Candidate B gains 1 mark for referring to proteins in the membranes being denatured.
>
> A major limitation of this procedure is that the discs were not held at the target temperatures (30°C to 70°C). This means that the independent variable should be called 'initial temperature', rather than 'temperature', because the water will cool as soon as it is added to the discs. It is impossible for each candidate to have five separate water baths in a practical task, so cooling is inevitable. However, the damage to the membranes may occur in the first few minutes of immersion in the water so it may make no difference to the results. This could be investigated by comparing the results of this investigation with one in which the discs are kept at the target temperatures for 10 minutes. The higher temperatures may increase the rate of diffusion of betalain from the cells so that the colours are darker and the concentrations of betalain are higher.

> **Remember that Candidate A has full marks for the qualitative task. Candidate B gains 4 marks out of 10.**

AS quantitative task: the effect of enzyme concentration on enzyme activity

Background

The enzyme lipase catalyses the hydrolysis of triglycerides. Milk fat is used as a source of triglycerides in this investigation. It is not possible to see any change in the appearance of milk as triglycerides are hydrolysed; however, the pH of milk changes during the course of the reaction. Phenolphthalein is used to detect a change in pH because it is an indicator that changes colour between pH 8 and pH 10.

Introduction

For this practical task you will **investigate the effect of enzyme concentration on the rate of an enzyme-catalysed reaction**.

You are given full instructions for the practical procedure, which must be followed carefully. **Before you start any practical work, you are advised to read through the procedure carefully and study Table 2**.

It is your responsibility to organise your time efficiently and to work safely. You will be assessed on the accuracy of your results.

For this task you could be awarded 2 marks for demonstrating the following:
- skilful practice   (*2 marks*)

Materials

You are provided with the following:
- full-fat milk
- sodium carbonate (Na_2CO_3) solution
- 5% lipase solution
- phenolphthalein

Safety

The experiment must be carried out safely.

Take care when using the following solutions:

Phenolphthalein	Flammable	
Sodium carbonate	Irritant	✖

Part 1

Procedure

Read steps **1** to **12** and study **Table 2** before you start your practical work.

1 You are provided with six test tubes labelled **A** to **F**. Use one of the 10 cm^3 syringes to put 5 cm^3 of **milk** into each of the test tubes **A** to **F**.

2 Use the other 10 cm^3 syringe to put 5 cm^3 of the sodium carbonate solution into each of the test tubes **A** to **F**.

3 Use a pipette to add five drops of phenolphthalein to each of the test tubes **A** to **F**. Put the bung provided into test tube **A**. Invert and shake until the contents are a uniform pink colour. Repeat this procedure with test tubes **B** to **F**.

4 Stir the lipase solution with the glass rod provided. Use the 2 cm^3 syringe and the 1 cm^3 syringe to make up a range of concentrations of lipase solution in test tubes **1** to **6** as shown in **Table 2**.

Table 2

Test tube	Volume of 5% lipase solution/cm^3	Volume of distilled water/cm^3	Concentration of lipase solution/%
1	0.0	2.0	0
2	0.4	1.6	1
3	0.8	1.2	2
4	1.2	0.8	3
5	1.6	0.4	4
6	2.0	0.0	5

5 Put some warm water in a beaker to act as a water bath. The beaker should be about half-full. Adjust the temperature of the water to 38°C (±2°C). Maintain the temperature of the water bath throughout the procedure.

6 Place test tubes **A** to **F** in the water bath for **at least 5 minutes**.

Now read carefully instructions 7 to 12.

7 Add the contents of test tube **1** to test tube **A**. Mix the contents by putting a bung into test tube **A** and shaking vigorously for about 5 seconds. Return test tube **A** to the water bath.

8 Start the stopwatch or stop clock and immediately add the contents of test tube **6** to test tube **F**. Mix the contents as before and return to the water bath.
   In **Table 3,** record the time taken for the pink colour to disappear completely.

9 Repeat step **8** with the remaining pairs of test tubes:
   test tube **5** to test tube **E**
   test tube **4** to test tube **D**
   test tube **3** to test tube **C**
   test tube **2** to test tube **B**

In each case, record in Table 3 the time taken for the pink colour to disappear completely.

10 If there is no change in a test tube after 10 minutes, record this as 'no change'.

11 Observe the contents of test tube **A**.

12 Use test tubes provided labelled **7** to **11** and **G** to **K** to obtain replicate readings for 1% to 5% lipase concentrations.

Do not repeat 0% lipase. Record your results in **Table 3**.

Table 3

Concentration of lipase/%	Time taken for the pink colour to disappear/s	
	Replicate 1	Replicate 2
0		
1		
2		
3		
4		
5		

(*2 marks*)

Part 2

1 Using the data you recorded in **Table 3** in **Part 1** of this task, calculate the mean time taken for the pink colour to disappear for each concentration of lipase. Record the mean times in **Table 4**. (*1 mark*)

Table 4

Concentration of lipase/%	Mean time taken for pink colour to disappear/s	Relative rate of activity/s^{-1}
0		
1		
2		
3		
4		
5		

2 The relative rate of activity at each concentration of lipase is calculated as follows:

$$\text{relative rate of enzyme activity} = \frac{1000}{t}$$

where t = mean time in seconds

Using the formula, calculate the relative rate of activity at each concentration. **Write your answers in Table 4**. (*1 mark*)

3 Use the processed data in **Table 4** to draw a graph to show the effect of increasing the concentration of lipase on the relative rate of enzyme activity. (*3 marks*)

4 Use your graph to describe the effect of increasing the concentration of lipase on the relative rate of enzyme activity. *(1 mark)*

5 By using phenolphthalein you were able to follow the hydrolysis of triglycerides in milk.

Explain fully why sodium carbonate and phenolphthalein were included in each reaction mixture. *(2 marks)*

Total: 10 marks

Candidates' answers to the AS quantitative task: part 1

Candidate A

Concentration of lipase/%	Time taken for the pink colour to disappear/s	
	Replicate 1	Replicate 2
0	No change	
1	203	310
2	103	119
3	39	44
4	21	20
5	13	15

Candidate B

Concentration of lipase/%	Time taken for the pink colour to disappear/s	
	Replicate 1	Replicate 2
0	No change	
1	150	215
2	88	85
3	22	35
4	10	12
5	5	12

The topic is based on the factors that influence enzyme-catalysed reactions from Module 1 of Unit F212. You can read more about this on pp. 29 to 38 in the Unit Guide for F212.

Both candidates have recorded all times to the nearest second. It is not appropriate to record the timings to any greater precision than this even though the stopwatch gives times to the nearest 0.01 second. Both candidates worked carefully through the procedure using test tube A, which remained pink, as a comparison. Candidate A used the spare test tube provided to make up a dilute solution of milk to compare with the test tubes to help identify the end point.

The teacher gave 2 marks to each candidate even though Candidate B found little difference between the results for 4% and 5%.

Part 2

Candidate A

1 and 2

Concentration of lipase/%	Mean time taken for pink colour to disappear/s	Relative rate of activity/s^{-1}
0	0	0.0
1	257	3.9
2	111	9.0
3	42	23.8
4	21	47.6
5	14	71.4

Candidate B

1 and 2

Concentration of lipase/%	Mean time taken for pink colour to disappear/s	Relative rate of activity/s^{-1}
0	0.0	0.0
1	182.5	5.4
2	86.5	11.6
3	28.5	35.0
4	11.0	90.9
5	8.5	117.6

Candidate A has calculated the means to the nearest whole number and Candidate B has calculated them to 1 decimal place. Both these approaches are acceptable. Your teacher will check your calculations and you will lose a mark if they are incorrect. Both candidates have calculated the mean correctly. Candidate B has not rounded the rates of activity for 1% and 3% properly, so loses 1 mark. The teacher would give Candidate B credit for calculating the means correctly, but would not give a mark for the calculation of the relative rates. Candidate B gains 1 mark for Question 1 and Question 2.

practical tasks

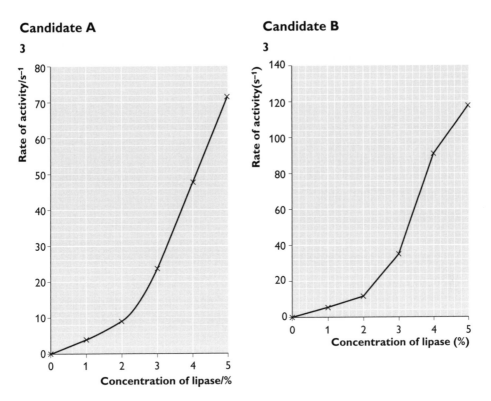

Candidate A

Candidate B

3 Both candidates have drawn line graphs and have scaled the axes correctly. The values of the independent and dependent variables (0 to 5% and 0 to $80\,s^{-1}$/0 to $120\,s^{-1}$) are placed at even intervals along the x- and y-axes respectively. Both axes are labelled with units and the units are separated from the variables. Candidate A has used a solidus (/) and B has used brackets. Good use of space with correctly presented axes gains 1 mark. Both candidates gain a mark for plotting the points correctly. Candidate A has drawn a curve of best fit and Candidate B has drawn straight lines between the points. These are both accepted for the third mark as neither candidate has extended the line beyond the last plotted point (5%). Candidate B gains 3 marks.

Candidate A

4 A longer time means a slower rate — rate is inversely proportional to the time taken. The relative rate of activity of lipase is proportional to its concentration. The rate increases steeply after 3%.

Candidate B

4 As the concentration of lipase increases so does the activity of lipase, until 4% when the relative rate begins to level off.

3 Both candidates have described the trend for their graphs. There is only 1 mark for this question so there is no requirement to give any figures. However, if

the trend was an increasing rate with a plateau, then figures would probably be expected for the concentration at which the rate becomes constant or where there is an inflection in the line (see p. 46). Candidate B gains 1 mark.

Candidate A

5 Sodium carbonate is alkaline and gives the milk a high pH above the range 8 to 10. At the start, when the phenolphthalein is added the colour is pink. As the pH decreases, phenolphthalein loses its pink colour and the normal colour of milk reappears. This gives the end point, which means there is a way of timing how long the reaction takes. The pH decreases because triglycerides are hydrolysed to fatty acids which neutralise the sodium carbonate.

Candidate B

5 The pH decreases as the triglycerides are catalysed. The pH decreases below 8 in this reaction so resulting in phenolphthalein returning to a colourless state.

Candidate A understands that it is necessary to raise the pH into the range in which phenolphthalein changes colour. Sodium carbonate is added to make the pH high enough for this at the beginning of the reaction. Candidate B has not conveyed this idea so does not gain a mark.

Candidate A is correct in saying that the pH decreases because triglyceride molecules are hydrolysed to form fatty acids. The fatty acids dissociate to give hydrogen ions that are responsible for neutralising the alkali in the milk, so decreasing the pH. As the pH *decreases* through the range 10 to 8, phenolphthalein is decolourised and the normal colour of milk can be seen. Candidate B does not show a good understanding of the reaction. 'Triglycerides are catalysed' is incorrect use of terminology. The candidate does not explain that they are hydrolysed to fatty acids. When answering the questions always look back to the beginning of the question paper to check what it is you are investigating. You will also find some of the terminology appropriate to the task. Make sure that you use these terms in your answers.

Candidate B gains 7 marks out of 10 for the quantitative task.

AS evaluative task: the effect of enzyme concentration on enzyme activity

Student Experiment Sheet

A student carried out an investigation into the effect of lipase concentration on the rate of hydrolysis of fat. The student followed the same procedure as in **Part 1** of the **quantitative task**, with the exception of carrying out three replicates, rather than two. The student's results are in **Tables 5** and **6** and are shown in **Figure 1**.

Table 5

Concentration of lipase/%	Time for pink colour to disappear/s		
	Replicate 1	Replicate 2	Replicate 3
0	0	0	0
1	197	215	187
2	91	87	84
3	48	55	52
4	36	32	40
5	33	30	28

Table 6

Concentration of lipase/%	Relative rate of enzyme activity/s^{-1}			
	Replicate 1	Replicate 2	Replicate 3	Mean
0	0.0	0.0	0.0	0.0
1	5.1	4.7	5.3	5.0
2	11.0	11.5	11.9	11.5
3	20.8	18.2	19.2	19.4
4	27.8	31.3	25.0	28.0
5	30.3	33.3	35.7	33.1

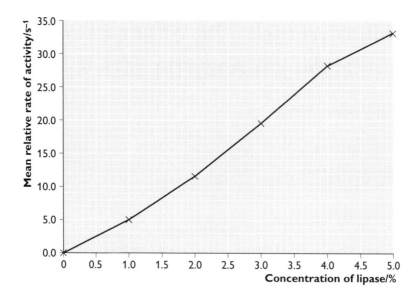

Figure 1

1 (a) Identify the following variables in this investigation:

independent variable, dependent variable, one controlled variable. *(3 marks)*

(b) The student included test tube **A** as part of the procedure. Why was this test tube included in the investigation? *(1 mark)*

2 Explain the effect of increasing the concentration of lipase on the relative rate of enzyme activity. *(3 marks)*

3 Predict **and** explain what you would expect to happen if the student used concentrations of lipase higher than 5%. *(1 marks)*

4 The student commented that it was not possible to give a **valid conclusion** about the effect of increasing the lipase concentration **up to 5%**. The student justified this by stating that it is not possible to be certain about the trend shown in **Figure 1**.

With reference to the student's results in **Tables 5** and **6**, explain why it is **not** possible to be certain about the trend shown in **Figure 1**. *(3 marks)*

5 The student used a 10 cm³ syringe to dispense the milk and the sodium carbonate solution.

Explain why you think this step in the procedure could be criticised for poor precision **and** explain how the **percentage error** of a volume dispensed by a 10 cm³ syringe is calculated. *(3 marks)*

6 State **two** major limitations of the method, **other than using syringes**, that may have affected the quality of the results obtained. For each limitation:

• explain how it influenced the quality of the results

• describe how you would modify the procedure to overcome the limitation

Be as specific as you can in describing your improvements. *(6 marks)*

Total: 20 marks

Candidates' answers to the AS evaluative task

Candidate A

1 (a) *IV* — concentration of lipase; *DV* — time for phenolphthalein to change colour; *CV* — concentration of milk fat

Candidate B

1 (a) *IV*—lipase; *DV* —colour change; *CV*—temperature

> Candidate B has identified neither the *concentration* of lipase as the independent variable nor the correct dependent variable, so gains only 1 mark.

Candidate A

(b) Test tube **A** was included as a control to show that there was no change in colour (or decrease in pH) in the absence of enzyme.

Candidate B

(b) This was included as a control to check that hydrolysis only occurs when an enzyme is present.

> Both candidates have explained why test tube **A** was included. If they had just said that it was a control, no mark would have been awarded. If you are asked a question like this, you must always explain how the tube acts as a control. Candidate B gains 1 mark.

Candidate A

2 Triglycerides in the milk are the substrate for the lipase enzyme. As the concentration of lipase increases there are more collisions between enzyme molecules and substrate molecules. Triglyceride molecules fit into the active sites of the lipase molecules to form enzyme-substrate complexes and from 1% to 5% there are more of these formed in the reactions. Throughout the range of concentrations, lipase is the limiting factor.

Candidate B

2 The graph shows an expected curve with a positive correlation between the enzyme concentration increasing and the rate of enzyme activity increasing. The rate increases slowly at first and then much faster. At 1% the rate of activity is 5, but at 4% it is 28. This occurs because of the 'collision theory' as there is an increase in enzymes making the enzyme and substrate more likely to collide. More collisions results in more substrate being broken down.

> This question is asking about the theory of enzyme activity. However, Candidate B has misunderstood the question and given a description of the relationship from the graph. Candidate A has made good use of terminology about enzymes from Unit F212. Candidate B has explained that there is an increase in 'enzymes' making collisions more likely. This is just enough to gain a mark. The examiners

expect candidates at AS to refer to the concentration of lipase rather than just refer to 'enzymes'. Candidate B gains 1 mark.

Candidate A

3   Proportionality would still be seen until the substrate concentration becomes the limiting factor. At this point, the rate becomes constant because there is insufficient substrate to fill the additional active sites. Figure 1 shows that rate is increasing and shows no evidence of reaching a plateau. From the graph I cannot predict when this will happen.

Candidate B

3   The rate would level off because as I found when I drew my graph (in the quantitative task) the rate starts to decrease after 4% lipase.

> 🖉 Candidate B refers to the graph from the quantitative task. You do not have your data or written answers from the quantitative task when you take the evaluative task. You may wish to refer to your results, if you can remember them, but it is not expected that you will do this.
>
> When investigating the effect of enzyme concentration, the substrate concentration should always be in excess. This gives the straight-line relationship that you can see in Candidate A's graph in the quantitative task (p. 95. However, at some point the substrate concentration will become the limiting factor — there will not be enough substrate to fill the additional active sites. At this point the rate becomes constant. However, it does not *decrease* as maintained by Candidate B. This would mean that the rate reaches a peak at 4%, which it does not. Take care when describing relationships like that shown in Candidate B's graph (p. 95). The rate continues to increase after 4% and may become constant at a concentration greater than 5%; it does not decrease after 4%. In such descriptions, it is always best to include some figures from the graph.
>
> Candidate A says that it is not possible to predict *when* the rate will become constant. It is more correct to say that it is not possible to predict the lipase concentration at which this will happen. Take care to refer to the independent variable when describing trends or making predictions. 'When' implies time and the independent variable is not time in this graph. Candidate B does not gain a mark.

Candidate A

4   The results in Tables 5 and 6 show that there was quite a variation in the times recorded. There are only three replicates so it is not possible to be confident about the mean. It is possible that some of the results are anomalous and would not be used in calculating the mean if there were more replicates. Range bars plotted on Figure 1 would show how much variation there is in the results. This would give a band across the graph and the 'true' line could be anywhere within that band. The student does not have enough readings for each value of the lipase

concentration to be certain. There are only six points on the graph — results for intermediate concentrations would help to increase certainty. Rates of activity above 5% may not continue the relationship, although that may be because substrate concentration is becoming a limiting factor.

Candidate B

4   The results are not reliable because they were not repeated enough times. There is some uncertainty in the value of the means for rates of activity. The results are not precise because, in the quantitative task, when the lipase solution is added to the milk and sodium carbonate solution I could not start the stopwatch immediately because I had to insert the bung and invert the test tube and then place it back into the water bath and start the stopwatch. The timings are subjective as we had to decide when the colour had changed and sometimes it was hard to tell when this first happened. I think the results are inaccurate because the lipase would have already started to catalyse the reaction and I would not have included this in the time taken. I might also have left the clock running too long at the end not realising when the colour had disappeared.

Both candidates make good points here. Both refer to the reliability of the results in the sense of repeatability. Candidate A uses terminology appropriate for this, such as 'confident', 'range bars', 'anomalous', 'intermediate' and 'value' as in *value for the independent variable*. Candidate B says that there is some uncertainty in the value of the mean but does not explain that this could be quantified by calculating the standard deviation (although that would be better with more replicates). The point about the precision of timing is a good one in this context. When using the terms *accurate* or *inaccurate* in an evaluation the candidate is referring to the 'true value'. This is the correct use of the term, but the candidate could have gone on to explain that it is difficult with this method of timing to be sure of the true values for rates. The subjective nature of the end point is a good comment; this may account for the variation in the results. Candidate B gains 2 marks.

Candidate A

5   It is difficult to line up the plunger and the graduations on a syringe, especially a 10 cm³ syringe. The percentage error is calculated by taking the volume that is halfway between graduations and expressing it as a percentage of the total volume being dispensed. So if a 10 cm³ syringe is used to dispense 10 cm³ the percentage error is:

$$\frac{0.5}{10} \times 100 = 5\%$$

Candidate B

5   When you use a syringe there is always some liquid left in the nozzle that you can't get out. The plunger sometimes sticks so you can't get the right volume.

There is also a chance that air bubbles are left in the syringe. This is why the syringe is not very precise.

 Candidate B is wrong in saying that the liquid left in the nozzle is a problem with a syringe. This volume is accounted for in the volume dispensed by the syringe. A volume such as 4.5 cm³ cannot be dispensed accurately with a 10 cm³ syringe; you could use a 10 cm³ syringe to dispense 4 cm³ and a 1 cm³ to dispense the remaining 0.5 cm³. There are fine graduations on a 1 cm³ syringe at every 0.01 cm³ mark so the percentage error is much less. Plungers can be sticky. If this is the case, take the plunger out of the syringe, wet the end, put it back into the barrel and slide it in and out several times. 'Sticky plunger' is an excuse, not a valid answer. Air bubbles should be removed from the syringe by pushing the plunger right into the syringe barrel before placing it into the liquid, keeping the nozzle of the syringe under the surface and pulling out and pushing in the plunger several times before filling the syringe beyond the volume required. Then hold the syringe pointing upwards, tap it with a finger and expel gently any air bubbles and the excess liquid. Read the volume at eye level making sure the black ring of the plunger is at the appropriate graduation. Candidate B does not gain any marks.

Candidate A

6 Limitation 1 It is difficult to tell exactly when the end point occurs. The student may not have stopped the stop watch when the pink colour just disappeared. The student may not have always used the same end point as when I did the quantitative task I noticed that the colour did not always fade in the same way each time. This would make the results inconsistent. To improve this several trial runs of the procedure should be carried out; 'real' results should not be taken until confident that the same end point is used each time.

Limitation 2 After the lipase solution was added to the milk the mixture was inverted, but was not mixed again during the reaction. This means that some reactions might have happened faster or slower than they should have done. I would stir the milk and lipase at set intervals while doing the timing.

Candidate B

6 Limitation 1 It is hard to keep the temperature constant in the water bath and the lipase was not put in the water bath before adding it to the milk. Temperature affects how enzymes work. I would put the test tubes in a thermostatically controlled water bath and use a thermometer to check that the temperature is constant at 38°C.

Limitation 2 There were not enough concentrations of lipase. There should have been concentrations at 0.5% intervals so it would be more certain where to plot the line.

 Candidate A recognises two problems with a procedure that involves timing to a certain colour change:

- It is not always possible to be certain of the point when the colour change occurs.
- It is hard to apply consistently each time.

It is good practice to take one or more trial readings first so that you are sure of the end point. This is an improvement you could suggest if you give the difficulty in deciding when the reaction has reached the end point as a limitation. If you are studying chemistry you will have done this when carrying out titrations.

Candidate B states that temperature was not controlled very well, but has not explained the effect that this might have. 'Temperature affects how enzyme works' is not detailed enough for AS. The candidate should state that temperature influences the *rate* of the enzyme reaction and if it was below 38°C (which is likely) then the rate for some of the tests may have been slower than expected. That would make the results for timing longer than at 38°C. Conversely if the temperature of the water bath was taken to above 38°C the timings would be too short. If the timing is too long, then the rate becomes too small; if the timing is too short, then the rate becomes too fast. Candidate B's second limitation points out that there were not enough *intermediate* concentrations within the range of 0 to 5%. The answer implies that more should be taken at 0.5% intervals and suggests that this will give more confidence in plotting the line. Another point to note is that the lipase solutions were not equilibrated at the same temperature as the milk and sodium carbonate solutions. Note that you may be asked how the improvements you give would reduce the errors and improve the results. Candidate B scores 4 marks.

✍ **Candidate B gains 9 marks out of 20 for the evaluative task.**

Overall, Candidate B gains 20 marks out of 40 for Practical Skills 1. This may not be enough for an E grade.

You can see that Candidate B has lost marks for a number of different reasons.

Qualitative task

- **The table was not formatted correctly; the whole table should be enclosed by ruled lines.**
- **The headings to the columns were not informative.**
- **Units (°C and %) were omitted from the headings for the columns for temperature and concentration; they should either be given after a solidus (/) or in brackets.**
- **Units were given against all the temperatures in the table. Units should only be given in the headings.**
- **Conclusions were given in the table.**
- **The reasons for one of the steps in the procedure were not explained (Q.1).**
- **Theory about protein denaturation was not linked to the explanation for the loss of pigment from cells (Q.2).**

Quantitative task

- The calculations were not rounded up correctly.
- A key part of the procedure (the use of sodium carbonate) was not explained (Q.5).

Evaluative task

- Terminology that should be used in evaluating the procedure and data has not been used correctly.
- Two of the variables have not been identified correctly (Q.1 (a)).
- Theoretical knowledge from Unit F212 has not been used to explain the relationship between the independent and dependent variables (Q.2).
- Including irrelevant material in describing the graph when an explanation was required (Q.2).
- Imprecise use of language in describing the graph (Q.3)
- The problem with the use of the syringe has not been explained fully (Q.5).
- The effects of the limitations of the method on the data have not been explained (Q.6).

A good way to prepare for the evaluative task is to do a class exercise in which one student carries out a practical procedure while the rest of the class suggest limitations with the procedure, likely sources of systematic and random errors and problems with reliability and precision of the results. If you discuss this with others you should gain a good understanding of the likely points to make during the evaluative task.

F216: Practical Skills in Biology 2

A2 qualitative task: the activity of immobilised glucose isomerase

This task is based on learning outcomes from Units F215 and F212.

Background

High-fructose corn syrup (HFCS) is produced by hydrolysing corn (maize) starch to yield glucose and then processing the glucose to produce a high percentage of fructose. This industrial process involves the use of three different enzymes. First, starch is treated with amylase to produce short chains of sugars. Second, the enzyme glucoamylase breaks the chains down even further to give glucose. Third, the enzyme glucose isomerase converts glucose to a mixture of about 50% fructose and 50% glucose. Glucose isomerase is immobilised on resin beads and the sugar mixture is passed over it.

In this investigation you will follow the course of the reverse reaction — the conversion of fructose into glucose.

Introduction

For this practical task you will **investigate the effect of pH on the conversion of fructose to glucose by immobilised glucose isomerase.**

You are given full instructions for the practical procedure, which must be followed carefully. **Before you start any practical work, you are advised to read through the procedure carefully.**

It is your responsibility to organise your time efficiently and to work safely. You will be assessed on the accuracy of your results.

For this task you could be awarded 1 mark for demonstrating each of the following:
- safe working
- skilful practice

Materials

You are provided with the following:
- glucose isomerase immobilised on resin beads
- fructose solution at pH 5.0
- fructose solution at pH 7.0
- fructose solution at pH 9.0
- Diastix® test strips and a colour chart
- distilled water

Follow this procedure when using Diastix® test strips to test for glucose:
- Dip a strip into the liquid to be tested and remove immediately.
- Shake off any liquid that remains attached to the coloured strip.
- Place on a white tile and start a stopwatch or stop clock.
- After 30 seconds, match the colour of the test strip with the colour chart and note the glucose concentration in $g\,100\,cm^{-3}$.
- Ignore any colour changes that occur after 30 seconds.

Safety

The experiment must be carried out safely.

Part 1

Procedure

1 Using hot and cold water, prepare a water bath in a $250\,cm^3$ beaker. The temperature must be between 50°C and 65°C and must be maintained within this range during the investigation.

2 Test the fructose solution at pH 7.0 with a Diastix® test strip. Observe the colour; use the colour chart and record the concentration of glucose.

3 Carefully pipette out the water from the test tube leaving the immobilised enzyme at the bottom. Discard the water into the plastic beaker (or other container).

4 Use the $10\,cm^3$ syringe to put $10\,cm^3$ of fructose at pH 7.0 into the test tube with the enzyme.

5 Put the test tube into the water bath, note the time and take the temperature of the water bath again.

6 After 5 minutes remove a small sample from the test tube with a pipette and put it into a depression on a spotting tile. Do not remove any of the beads, only the solution. Allow 30 seconds for the solution to cool and then test it with another Diastix® test strip.
Observe the colour; use the colour chart and record the concentration of glucose.

7 Shake the test tube and leave for another 5 minutes.
While you are waiting you may wish to construct a table to record all your observations and results in this task.

8 After 5 minutes repeat steps **6** and **7.**

9 Leave the tube for another 5 minutes and repeat steps **6** and **7**.

10 After the final test, remove the test tube from its water bath. Pipette off the solution and place the liquid in the plastic beaker.

11 Wash out the syringe. Use the syringe to add $10\,cm^3$ of distilled water to the enzyme in the test tube. Shake and allow the beads to settle at the bottom of the tube.

12 Carefully pipette off the water into the plastic beaker and repeat the washing with another $10\,cm^3$ of distilled water.

13 Pipette off this water and then add $10\,cm^3$ of fructose solution at pH 5. Repeat steps **5** to **11**.

14 Repeat the whole procedure with fructose solution at pH 9.
Construct and complete a table to record your observations and results. (*3 marks*)

Part 2

1 Benedict's solution gives a positive result with glucose. Explain why Diastix®
test strips and **not** Benedict's solution were used to test for glucose in this
investigation. (*1 mark*)
2 Explain the results that you have obtained. (*3 marks*)
3 State **one** advantage of using immobilised glucose isomerase in the production
of high-fructose corn syrup. (*1 mark*)

Total: 10 marks

Candidates' answers to the A2 qualitative task: part 1

✍ Candidate A kept all the temperatures within the range specified, Candidate B
did not raise the temperature when it reached the bottom of the range so lost
the mark for skilful practice. Both candidates worked tidily and safely, wearing
eye protection when heating the water bath. Candidate B gains 1 mark.

Candidate A

pH	Initial temp. /°C	Colour of Diastix (and concentration of glucose/$g\,dm^{-3}$)		
		Time/min		
		1	6	11
5	65	Blue (0)	Blue-green (0.10)	Green (0.25)
7	61	Blue (0)	Blue-green (0.10)	Green (0.25)
9	62	Blue-green (0.10)	Green (0.25)	Light brown (0.50)

Candidate B

pH	Colour of Diastix at minute intervals		
	1	5	10
5	Blue	Blue-green	Green
7	Blue	Blue-green	Green
9	Light-green	Green	Light brown

pH	Concentration of glucose in $g\,dm^{-3}$ at minute intervals		
	1	5	10
5	0	0.1	0.25
7	0	0.1	0.25
9	0.10	0.25	0.50

The candidates have similar results. Candidate A presents both the colours of the Diastix® and the concentrations of glucose in the same table. Candidate B uses two tables — one for the colours and one for the concentrations. The task asks for a table to record observations and concentrations so there should be a single table. Candidate B's headings for the columns are muddled. The unit for concentration should be in brackets or separated from 'concentration of glucose' by a solidus. It would be best to show the time on a separate line as Candidate A has done. 'At minute intervals' suggests at intervals of 1 minute. Candidate B has not noticed that there were also 5-minute intervals after the first sample, so giving times of 6 and 11 minutes. Candidate B fails to score.

Part 2

Candidate A

1 Benedict's solution would not be of any use in this investigation because it tests for reducing sugars. Both glucose and fructose are reducing sugars. All the samples tested would be positive. Diastix is specific for glucose as it gave a negative result for fructose.

Candidate B

1 It is easy to use the Diastix test strips as you do not need to have a boiling water bath as you do for the Benedict's test. It is easy to match the colours from the test strips to the colour card and then read off the concentration of glucose.

Candidate A has explained in full why Benedict's solution would not be of any use in this investigation. Candidate B has answered a different question by explaining how much easier it is to use test strips rather than Benedict's solution. Candidate B fails to score.

Candidate A

2 Glucose isomerase converts fructose to glucose. This is a reversible reaction that occurs in glycolysis. The rate of conversion is fastest in the solution at pH 9. This may be the optimum pH, but with only a limited number of buffer solutions used we cannot be sure. This pH favours either the formation of enzyme-substrate complexes or has an effect on the way in which the enzyme is immobilised on the beads. This may happen because the hydrogen bonds and ionic bonds that contribute to the tertiary structure of enzyme molecules break and the active site changes shape. This means that the shape changes so the substrate (fructose) does not fit so well. More bonds that hold enzyme molecules to the surface of the bead may form at lower pH so that active sites are not as accessible to the fructose molecules.

Candidate B

2 The beads contain enzymes that are affected by pH. The reaction was fastest at pH 9. pH affects the shape of the active site. There are a variety of bonds within the polypeptide chain that makes up the enzyme. These bonds hold the chain in a certain 3-D shape that is complementary to the substrate, fructose. The fructose molecules may fit more easily into the active site at pH 9, so the reaction to change it into glucose is faster than at other pH values.

Both candidates refer to information about active sites from AS. This highlights the fact that the A2 tasks are synoptic and you are expected to use information from the AS units. The suggestions made by both candidates are acceptable explanations. Changes in pH can influence the attachments of the enzyme to the beads and the shape of the active site. Candidate B gains 3 marks.

Candidate A

3   The beads of immobilised enzyme can be reused easily. They can also be put in a column and used over a long time.

Candidate B

3   The enzymes can be extracted from the reaction mixture by sieving or filtering. This is not possible with enzymes in solution.

Both answers are suitable. Candidate B gains 1 mark.

Candidate B gains 5 marks out of 10 for the A2 qualitative task.

You might be expected to compare the activity of immobilised and non-immobilised enzymes. They behave differently in different conditions as you can see in Figure 2.

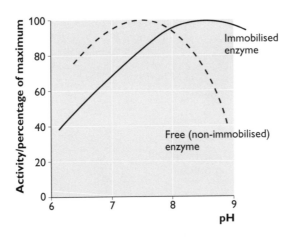

Figure 2

A2 quantitative task: the effect of temperature on the rate of respiration in yeast

Background

Yeast is a single-celled organism that respires both aerobically and anaerobically, releasing carbon dioxide.

Introduction

For this practical task you will **investigate the rate of respiration at different temperatures**. You are expected to choose at least five different temperatures (known as target temperatures in this task). However, the actual temperatures you use may vary by ±2°C from your target temperatures.

You are given full instructions for the practical procedure, which must be followed carefully.

Before you start any practical work, you are advised to read through the procedure and study Figure 3 and Table 7.

It is your responsibility to organise your time efficiently and to work safely. You will be assessed on the accuracy of your results.

For this task you could be awarded 2 marks for demonstrating the following:
- skilful practice

Materials

You are provided with the following:

boiling tube containing a suspension of yeast in a solution of glucose (do **not** shake this tube at any stage during the investigation)
- coloured liquid for the manometer
- supply of warm water (20°C)
- supply of warmer water for adjusting the temperature of the water bath

The manometer consists of capillary tubing ending in a reservoir (see **Figure 3**). When the three-way tap is in the closed position, moving the syringe plunger causes the coloured liquid to move up and down the capillary tubing.

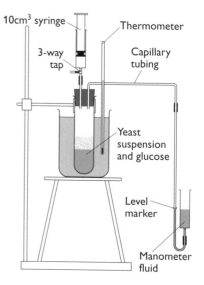

Figure 3

practical tasks

Safety

The experiment must be carried out safely.

Part 1

1 Set up a clamp stand, clamp, tripod, gauze and water bath as in **Figure 3.** Clamp the boiling tube securely so it fits inside the water bath.
2 Fit the manometer and syringe assembly as shown in the diagram ensuring that:
  - the bung makes an airtight seal in the mouth of the test tube
  - the three-way tap is open (turned upwards)
3 Use a pipette to fill the reservoir of the manometer about three-quarters full of coloured liquid. If the liquid does not flow into the capillary, close the tap (horizontal) and raise the syringe plunger carefully. Air bubbles in the column can be expelled by using the syringe to push the liquid and trapped air back to the reservoir.
4 Turn the tap downwards, push the syringe plunger to the bottom of the barrel and open the tap (vertically upwards).
Move the rubber ring on the capillary tube to indicate the level of liquid.
5 Record the temperature of the water bath in the first row in **Table 7**. Start a stopwatch or stop clock and close the tap (horizontal). Leave the tap closed for 1 minute and keep withdrawing the syringe plunger to bring the liquid back to the marker. After 1 minute bring the liquid level exactly back to the mark, open the tap (upwards) and record the volume indicated on the syringe.
6 Turn the tap downwards, push the syringe plunger to the bottom and close the tap (horizontal).
7 Take two more readings of the volume of gas produced in 1 minute. Record the temperature each time.
8 **You are going to increase the temperature of the water bath in intervals of 5°C**. The temperatures that you intend to use are **target temperatures**. Write down the six target temperatures you intend to use in the first column of **Table 7**.

Table 7

Target temperature /°C	Actual temperature /°C	Volume of carbon dioxide/cm^3

9 Check that the tap is open (upwards) and use some warm water to raise the temperature of the water bath by about 5°C. Leave the apparatus for 2 minutes after heating has stopped. Record the new temperature and take three more 1-minute readings as before, recording the temperature at the start of each reading.

10 Repeat the operation from step **8** by raising the temperature about 5°C each time until you have results for five or six different temperatures. Take the readings as quickly as you can.

11 Record all your results in **Table 7**.

(2 marks)

Part 2

1 Using the data you recorded in **Table 7** in **Part 1** of this task, calculate the mean temperature and the mean volume of carbon dioxide produced for each of your target temperatures. Record the mean temperatures and mean volumes in **Table 8**.

(1 mark)

Table 8

Mean temperature/°C	Mean volume of carbon dioxide/ cm³

2 Use the data from **Table 8** to plot a graph of the rate of respiration against temperature.
(3 marks)

3 The temperature coefficient (Q_{10}) is the relative change in the rate of a process with an increase in temperature of 10°C.
Use your graph to estimate the Q_{10} for any part of the temperature range that you investigated.
State the temperature range and your estimate of Q_{10}. *(2 marks)*

4 Use your data to describe how temperature influences the rate of respiration in yeast. *(2 marks)*

Total: 10 marks

Candidates' answers to the A2 quantitative task: part 1

Candidate A

Target temperature/ °C	Actual temperature/ °C	Volume of carbon dioxide/cm^3
20	20	0.3
	18	0.4
	18	0.3
25	26	0.7
	25	0.9
	25	0.3
30	32	1.3
	31	1.5
	30	1.7
35	37	3.6
	36	3.4
	36	3.8
40	41	5.4
	40	5.3
	38	5.2
45	45	7.3
	45	7.2
	43	7.8

Candidate B

Target temperature/ °C	Actual temperature/ °C	Volume of carbon dioxide/cm^3
20	22	0.4
	20	0.7
	19	0.6
25	24	1.3
	23	1.5
	21	0.7
30	33	2.7
	32	2.6
	29	2.4
35	38	4.7
	37	4.9
	36	5.3
40	42	6.3
	41	6.4
	38	6.6
50	49	3.6
	48	4.2
	48	3.8

Both candidates worked safely, carefully and followed the instructions about using the three-way tap correctly. Candidate A has kept all the temperatures within ±2°C of the target temperatures. Candidate B has not increased the target temperatures by 5°C each time and has not kept the actual temperatures within the ±2°C limits. However, Candidate B has collected a full set of results for six different temperatures so gains 1 mark.

Part 2

Candidate A		Candidate B	

1

Mean temperature/°C	Mean volume of carbon dioxide/ cm^3
18.7	0.3
25.3	0.6
31.0	1.5
36.3	3.6
39.7	5.3
44.3	7.4

1

Mean temperature/°C	Mean volume of carbon dioxide/ cm^3
20.3	0.6
22.7	0.6
31.3	1.2
37.0	2.6
40.3	6.4
48.3	3.9

Both candidates have calculated the mean temperatures and the mean volumes correctly. Notice that both have rounded up their calculations correctly. Candidate B gains 1 mark.

Candidate A

2

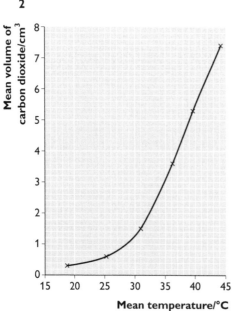

Candidate B

2

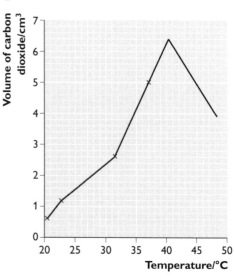

Both candidates have presented acceptable graphs. Candidate A has drawn a curve of best fit and Candidate B has joined the points with ruled lines. The rate of respiration as measured by the volume of carbon dioxide increases throughout the temperature range investigated by Candidate A. Perhaps because Candidate B took the temperature to nearly 50°C the rate reaches a peak at just over 40°C. There may be several reasons for this. Both candidates have a displaced origin for the independent variable. Candidate B gains 3 marks.

Candidate A

3 Temperature range 30–40°C

Estimate of Q_{10} = 6.4

Candidate B

3 Temperature range 20–30°C

Estimate of Q_{10} = 2

Candidate A has a Q_{10} value much higher than the value of 2 that we might expect. However, the readings are taken from the graph correctly. Candidate B has taken figures from the graph for the actual temperatures of 20.3°C and 31.3°C. As it happens, the figures are so close to what they would be for 20°C and 30°C that the examiner would award the mark.

Candidate A

4 The rate of respiration does not increase much between 18°C and 25°C. There is then an increase in the rate which becomes very steep after 30°C. The highest rate is reached at 7.4 cm^3 min^{-1} at 44°C.

Candidate B

4 The rate of respiration increases from about 20°C to then reach a peak at 40.3°C. It decreases to 3.9 cm^3 at 48.3°C.

Candidate B has an easier task describing the results as there is a peak in the rate of respiration at about 40°C. Both candidates have used the figures from their table and graph to illustrate their answers, but only Candidate A has given the correct unit for rate. Candidate B has referred to the rate but has omitted 'min^{-1}'. Candidate B gains 1 mark.

Candidate B gains 7 marks out of 10 for the quantitative task.

A2 evaluative task: the effect of ethanol on the rate of respiration in yeast

Student experiment sheet

Ethanol is a product of anaerobic respiration in yeast. A student carried out an investigation into the effect of increasing concentration of ethanol on the rate of respiration of yeast using the same materials and apparatus as in the quantitative task. The student used six different concentrations of ethanol making them up from a 40% concentration as shown in **Table 9**. The student then mixed equal volumes of a suspension of yeast in glucose with each ethanol concentration and then used the same apparatus (see Figure 3 on p. 110) and procedure as in **Part 1** of the **Quantitative task**, but carried out ten replicates at each concentration. The procedure was carried out at 30°C. The student's results are given in **Tables 10** and **11** and are shown in **Figure 4**.

Table 9

Volume of 40% ethanol/cm^3	Volume of distilled water/cm^3	Working concentration of ethanol/%	Final concentration when added to yeast and glucose/%
0.00	10.00	0.00	0.00
1.25	8.75	5.00	2.50
2.50	7.50	10.00	5.00
3.75	6.25	15.00	7.50
5.00	5.00	20.00	10.00
10.00	0.00	40.00	20.00

Table 10

Concentration of ethanol/%	Rate of respiration/cm^3 min^{-1}									
	Replicates									
	1	2	3	4	5	6	7	8	9	10
0.0	0.75	0.82	0.73	0.62	0.78	0.69	0.70	0.85	0.74	0.78
2.5	0.71	0.65	0.63	0.71	0.62	0.65	0.59	0.77	0.62	0.78
5.0	0.59	0.63	0.59	0.57	0.65	0.58	0.67	0.58	0.58	0.60
7.5	0.51	0.53	0.55	0.49	0.54	0.55	0.52	0.50	0.53	0.55
10.0	0.13	0.14	0.21	0.12	0.15	0.14	0.18	0.12	0.12	0.14
15.0	0.21	0.15	0.32	0.21	0.22	0.24	0.19	0.18	0.17	0.18
20.0	0.01	0.05	0.10	0.06	0.07	0.02	0.02	0.04	0.01	0.00

Table 11

Concentration of ethanol/%	Mean rate of respiration/cm^3 min^{-1}	Range/cm^3 min^{-1}	Standard deviation/ cm^3 min^{-1}
0.0	0.75	0.62–0.85	0.07
2.5	0.67	0.59–0.78	0.07
5.0	0.60	0.57–0.67	0.03
7.5	0.53	0.49–0.55	0.02
10.0	0.15	0.12–0.21	0.03
15.0	0.21	0.15–0.32	0.05
20.0	0.04	0.00–0.10	0.03

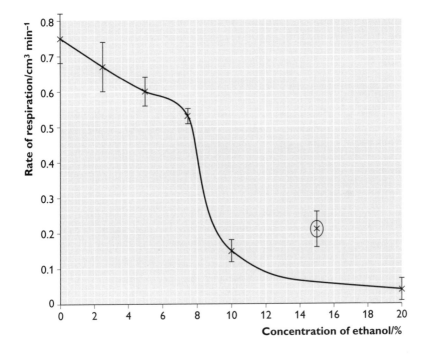

Figure 4

1 Explain why the student has ringed the plot for 15% ethanol on **Figure 4**.
(2 marks)

2 Describe **and** explain the relationship shown in **Figure 4**. *(4 marks)*

3 **Table 11** shows the standard deviations for the student's data.

   (a) Explain why calculating the standard deviation is more useful than calculating the range. *(1 mark)*

   (b) State how the student has used the standard deviations. *(1 mark)*

4 State **two** factors that may make it difficult for someone else to reproduce the results that the student obtained. *(2 marks)*

5 State **two** limitations of the method that may have affected the quality of the results obtained. For each limitation:
- explain how it influenced the quality of the results
- describe how you would modify the procedure to overcome the limitation

Be as specific as you can in describing your improvements. (6 marks)

6 Comment on the confidence that the student may have in the conclusion that can be drawn from the results. (4 marks)

Total: 20 marks

Candidates' answers to the A2 evaluative task

Candidate A

1 The result for 15% is anomalous as it does not fit the overall trend of a decrease in rate over this part of the concentration range. The expected result should be between the rates for 10% and 20%, but not much higher than that for 10%.

Candidate B

1 The result is an anomaly.

> The question asks for an explanation, so simply stating that the result is an anomaly is not sufficient. Candidate A has explained why it is seen as anomalous. Candidate B gains 1 mark.

Candidate A

2 The ethanol does not have much effect on the rate of respiration of the yeast until a concentration of about 5% when the rate is $0.60\,cm^3\,min^{-1}$. Between 7.5% and 10%, there is a steep decrease in the rate of respiration from $0.53\,cm^3\,min^{-1}$ to $0.15\,cm^3\,min^{-1}$. At high concentrations (20%) there is virtually no respiration. Ethanol at this concentration must be toxic to the yeast cells. Ethanol is an organic solvent so it probably dissolves the membranes of the yeast cells and it may dry out the cells because ethanol is hygroscopic and will absorb water from the cells. This may cause denaturation of proteins, which could include membrane proteins and the enzymes involved in respiration. Ethanol may also act directly on the enzymes — possibly by being either a competitive or a non-competitive inhibitor.

Candidate B

2 The rate of respiration decreases between 0% and 20% ethanol. The steepest decrease is in the middle of the range. The rate of respiration is highest at $0.75\,cm^3\,min^{-1}$ and this decreases to $0.02\,cm^3\,min^{-1}$. The ethanol inhibits respiration. Yeast cells cannot survive in high concentrations of ethanol which is produced in anaerobic respiration. When yeast cells respire anaerobically, pyruvate is converted into ethanol and carbon dioxide. It is the carbon dioxide that has been measured in the experiment.

🖉 Candidate B's description of the results concentrates on the two extremes of the range. Figures are quoted, but again for the extremes of the range. It is not enough to refer to a steep decrease in the middle of the range, the ethanol concentrations should be identified and some more figures given. Candidate A uses knowledge from Units F211 and F212 to explain why the rate of respiration decreases. Candidate B does not offer enough detail to gain any marks for the explanation, but gains 1 mark for the description of the overall effect of ethanol.

Candidate A

3 **(a)** The standard deviation shows the spread of results about the mean. We can be 95% certain that the mean lies within 2 standard deviations of the calculated mean. This gives a 95% confidence level so the student can say that the true mean lies within ±2 SD of the means given.

(b) The standard deviations have been plotted on the graph as error bars.

Candidate B

3 **(a)** The advantage of calculating standard deviations is that it shows how the results are dispersed about the mean.

(b) They show how much variation there is about the mean so we can make valid comparisons between the rates.

🖉 In part **(a)**, Candidate B makes a statement that could be applied to the range. It is better to be specific here: 68% of the readings are within ±1 SD of the mean; 95% are within ±2 SD of the mean. The range may include outliers that are not representative of the results as a whole. Candidate B has obviously not spotted the error bars on Figure 4. Always look at the SES as you plan each answer to a question. There may be useful information that you can use or that will act as a clue to what you should write. Candidate B has not given precise answers and so fails to score.

Candidate A

4 When yeast respires anaerobically it produces ethanol. If yeast produces ethanol in the experiment then this will increase its concentration. This might be more important at the lower concentrations. We do not know if yeast respires aerobically or anaerobically in this apparatus and someone carrying out the task needs to know how long to leave the mixtures before taking results. The student has not stated how long the yeast suspension was left to mix with the different ethanol concentrations before the results were taken. The effect of ethanol on yeast may not be immediate. Yeast may be more able to tolerate ethanol the longer it is exposed to it.

Candidate B

4 It would be difficult to reproduce the results because the concentrations of yeast and glucose are not given. The yeast suspension was already respiring

when I started the task. I do not know how long it had been mixed together or how much glucose had been used up already. Maybe the glucose concentration was becoming a limiting factor. I took all my readings from the same mixture. The replicates in the table may have been taken from one reaction mixture per concentration or from two reaction mixtures — it is not clear.

📝 Remember that reproducibility is one aspect of the reliability of results. Candidate A refers to the biology of yeast and uses knowledge of aerobic and anaerobic respiration from Module 4 in F214. The time interval between mixing yeast and glucose and then adding ethanol is not given and Candidate A has done well to identify this. Candidate B has referred to the quantitative task, rather than the investigation given in the SES ('.... when I started the task.') However, the points made are appropriate to the task in the SES so Candidate B gains 2 marks. Read the questions in this task very carefully – they refer to the information given in the SES. Of course, you should use your experience of the quantitative task if that is relevant to the question asked.

Candidate A

5 **Limitation 1** The conditions changed in the test tube while I was taking results and it would be the same in this task. The yeast would have been absorbing and respiring the substrate, glucose, so its concentration would decrease continuously. The results, especially at the low concentrations of ethanol, may all be less than they should be if glucose is not a limiting factor. I would set up the ten separate suspensions for each concentration at intervals and leave them for 10 minutes at 30°C and then record the carbon dioxide produced.

Limitation 2 The volume of carbon dioxide collected was small. This means that the percentage error is quite large so there could be quite a variation in results. It is hard to say how accurate the results are. I would overcome this by leaving the apparatus for longer and collect $10\,cm^3$ carbon dioxide each time. I would use $10\,cm^3$ as this is the volume of the syringe.

Candidate B

5 **Limitation 1** During the time when I was taking each replicate the ethanol concentration would be increasing if the respiration is anaerobic. I don't know if the respiration is anaerobic, but in the quantitative task we were told not to shake the tube with the yeast suspension. If the tube was shaken then some oxygen would dissolve in the suspension and some aerobic respiration would occur (oxygen is not very soluble in water). My improvement would be to do a trial experiment to see how much, if any, ethanol is produced. I could try a different method to measure the rate of respiration. I could bubble oxygen through the suspension and see how much glucose is used up. Diastix could be used to see how long it takes for all the glucose to be used.

Limitation 2 It is difficult to take readings with a $10\,cm^3$ syringe. It is difficult to pull up the syringe plunger to position the meniscus under the ring on the capillary tubing. It is only possible to measure to the nearest 0.5 cm³.   My improvement would be to use a gas syringe to collect the gas.

✒ Both candidates have given appropriate limitations and suitable improvements. Candidate A explains the effects that the limitations may have on the quality of the data. It is a common mistake to allow too short a time for organisms to adjust to changed conditions. If the yeast suspension was made from dried yeast, then it would take longer than 10 minutes to reach a constant respiration rate. Since you cannot record the rate of respiration from all the tubes at the same starting time, the candidate suggests starting them at intervals. This staggered start is a good practical procedure to suggest as an improvement. We do not know how long it takes yeast to reach a constant rate of respiration. Candidate B mentions trial experiments. You could find out how long it takes for yeast to reach a constant rate of respiration by setting up a respirometer and taking readings every 5 minutes for an hour. This would be similar to Example 23. Candidate B has not explained what effect the limitations would have on the quality of the data. Nevertheless, Candidate B gains 4 marks for this answer because the limitations and improvements given are good.

Candidate A

6 I would conclude from the results that an increase in ethanol concentration causes a decrease in the rate of respiration. The biggest decrease happens between 5% and 10%. However, I do not know what the exact relationship is. The results for 15% appear to be anomalous as they do not fit the trend on the graph. The error bars for 10% and 20% suggest that the 15% result is anomalous but it would have to be repeated to make sure. The results are fairly reliable as the replicates are close together and the error bars are small. More readings need to be taken at concentrations between 5% and 10% to identify the critical concentration at which the rate drops significantly. Since we don't know whether the concentration of glucose is also a limiting factor we don't know what effect this may have on the results. The results may not be accurate as the mixtures may have been left for different lengths of time before taking results and we also don't know how long it takes for ethanol to have an effect on yeast.

Candidate B

6 It is difficult to be confident in the conclusion that ethanol affects yeast. There are so many variables that could affect respiration in the investigation and there are not enough results. For example, there are not enough concentrations in the range 5% to 10% to be sure where to draw the line of best fit. We also do not know whether yeast has reached a constant rate of respiration when the results were taken. This would make the results not valid. I think that the results were not reliable, accurate or precise.

✒ Candidate A gives a well-argued answer to this question. The terms 'reliable', 'replicates', 'anomalous' and 'accurate' are all used correctly. In a question like this you are expected to summarise the effects of the systematic and random errors and the limitations to explain how confident you are in your conclusions. Both candidates discuss the problem of there not being enough intermediate readings to be certain of the relationship between 5% and 10%. This is something that could be determined and compared with other people (reproducibility) or checked by looking at scientific literature (accuracy — closeness to the true relationship). Candidate B also points out that we do not know whether the rate of respiration has reached a constant value, which is a valid point. Candidate B gains 2 marks. The last sentence includes three useful words — reliable, accurate and precise — but offers no further explanation.

Candidate B gains 10 marks out of 20 for the evaluative task.

Overall, Candidate B gains 22 marks out of 40. This may not be enough for an E grade.

You can see that Candidate B has lost marks for a number of different reasons.

Qualitative task

- Not recording the temperatures.
- Using two tables to record the results. You should always use one table for raw data. You may use a second table for recording derived variables, e.g. rates of reaction.
- Not answering the question, (Q.1).

Quantitative task

- The procedure in Part 1 has not been followed with regard to the temperatures.
- The correct unit for rate has not been used (Q.4).

Evaluative task

- No explanation is given for why a result is anomalous (Q.1).
- An incomplete description of the results on the SES is given (Q.2).
- No explanation for the results is given (Q.2).
- Answers do not include information from the SES (Q.3).
- The effects of the limitations on the data are not given (Q.5).
- No justification is given for stating that results are not reliable, accurate or precise (Q.6).

'Try this yourself' answers

1 A, oak; B, ash; C, beech; D, horse-chestnut; E, rowan; F, sycamore; G, plane; H, cherry; J, lime; K, silver birch. Note that ash and rowan have compound leaves with leaflets arranged opposite each other on a central axis. The leaves of ash are arranged opposite each other on the stem and the leaves of rowan are arranged alternately on the stem.

2 Quantitative features, such as mass, length, width, as these are so variable

3 $0.9\,g\,100\,cm^{-3}$; $9\,g\,dm^{-3}$

4

Volume of 2% NaCl solution/cm³	Volume of water/ cm³	Concentration of NaCl/%
0.0	10.0	0.0
1.5	8.5	0.3
3.5	6.5	0.7
4.5	5.5	0.9
7.5	2.5	1.5

5 Water moved out of the cells by osmosis down a water potential gradient; cell volume has decreased.

6 IV — potential inhibitors and sources of amylase; DV — presence or absence of starch.

7 As the control to show that, in the absence of an inhibitor, starch is hydrolysed under the conditions of the investigation.

8 Each of the four substances tested inhibited all four amylases.

9 Cell membranes become more permeable at high temperatures and chloride ions leak out of the cells; silver nitrate detects the presence of chloride ions. Silver chloride forms a cloudy suspension.

$Ag^+ (aq) + Cl^-(aq) \rightarrow AgCl(s)$

10 16% and 9%

11

Cube of side/ mm	Surface area of cube/mm²	Volume of cube/mm³	Surface area: volume ratio	Time for acid to diffuse to centre of cube/s		
				Replicate 1	Replicate 2	Mean
1	6	1	6:1	8	11	10
2	24	8	3:1	26	26	26
3	54	27	2:1	43	45	44
4	96	64	1.5:1	65	80	73
5	150	125	1.2:1	112	122	117

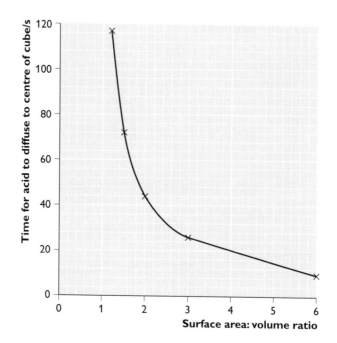

12 A large cube has a small surface area-to-volume ratio; graph shows that substances take a long time to diffuse to the centre; the same would happen in organisms; diffusion from the surface to the centre of the organism is too slow to supply oxygen. A specialised surface for gas exchange (e.g. lungs or gills) gives more surface for diffusion; a transport system distributes oxygen throughout the body.

13 Each piece was exposed to the same conditions; the potato pieces change the water potential of the sucrose solution and this could influence what happens in other pieces.

14

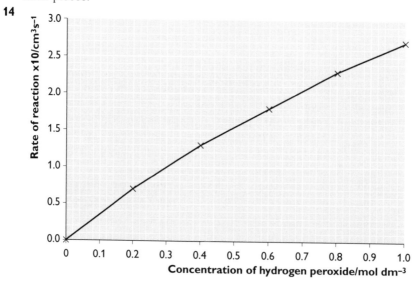

15 Increase in concentration of substrate molecules leads to an increase in the number of collisions between enzyme molecules and substrate molecules per unit time, so the reaction rate increases.

16

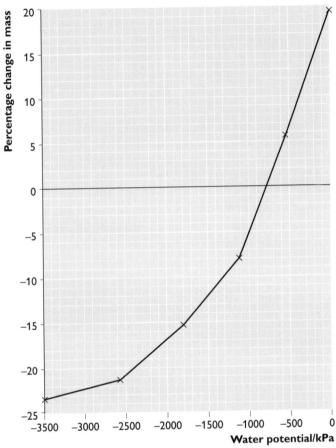

The water potential of the potato tissue = −800 kPa

17 They would be heavier than they should be.

18 The mass change will be underestimated so the water potential found will be lower than the true value.

19 A is a comparator to help identify the end point each time.
B is a control to show that DCPIP is not reduced in the light in the absence of chloroplasts.

20 Time how long it takes for the green colour to appear; rate = $1/t$.

21 Neutral red is absorbed by active transport (as none is left in the surrounding solution). The pH of the yeast cells is maintained at < 7 even though surrounding solution is alkaline. Boiling yeast cells kills them and shows that cells can no longer maintain a different pH from surroundings. Methylene blue is reduced and decolourised by living cells, but is oxidised when the solution is shaken. Reduction does not occur in dead cells showing the process is enzyme controlled.

22 The rate is calculated once it has become constant, so is taken from the straight line between 7 and 11 minutes. Rate = $0.1 \, cm^3 \, min^{-1}$.

23 Graph of oxygen uptake over time shows a constant rate over 270 s. Rate of oxygen uptake = $0.33 \, cm^3 \, min^{-1}$. Carbon dioxide production is calculated by subtracting the values for 'without soda lime' from the values for 'with soda lime'. This also shows a constant rate. Rate of carbon dioxide production = $0.23 \, cm^3 \, min^{-1}$.

24 Seeds may be respiring fats which are highly reduced molecules, so respiration produces much water and not as much carbon dioxide as when carbohydrate is respired.

25 Bar chart.

26 Some stored molecules are respired (see result for water); yeast does not respire lactose and galactose as results are similar to control (water); yeast respires glucose and fructose (monosaccharides) as these are small molecules and can diffuse across cell membrane; yeast has enzymes to hydrolyse maltose and sucrose and respires the monosaccharides released; yeast may not have enzymes for hydrolysis of lactose or isomerisation of galactose to glucose.

27 Cannot comment on repeatability as no student has carried out replicates. Reproducibility is poor as results vary considerably between students. Means of their results cannot be calculated. Student C has recorded to 0.01 s which is too precise for this investigation.

28 Treat the results separately. Convert to rate by calculating $1000/t$. Plot line graphs.

29 Rate of respiration increases to an optimum temperature and then decreases. Temperature influences collisions between enzyme molecules and substrate molecules and entry of glucose to the cells. At higher temperatures enzymes begin to denature.

30 A, urine; B, glomerular filtrate; C, blood plasma.

31 D, diabetes mellitus; E, nephritis; F, normal; G, diabetes insipidus; H, obstructive jaundice; J, starvation.

32 Damage to glomerulus or high blood pressure allows protein molecules into the filtrate. Concentration of glucose in the filtrate is above the renal threshold so not all can be reabsorbed in the proximal convoluted tubules.

33 11

34 14

35 $14/25 \times 100 = 56\%$

36 In the F_1 there should be a 1:1 ratio; in the F_2 there should be a 1:1:1:1 ratio.
Null hypothesis: there is no significant difference between the observed and expected numbers in the F_1 and F_2 offspring.
$F_1 \, \chi^2 = 2.1$; df = 1; $p > 0.05$ so difference in numbers is not significant
$F_2 \, \chi^2 = 3.9$; df = 3; $p > 0.05$ so difference in numbers is not significant

37 Plot line graphs of number of woodlice in each category against time. Investigation 1: with a null hypothesis that there is no significant difference between the numbers of woodlice in the two areas; a χ^2 test could be carried out on the final figures using 5 and 5 as the expected numbers in the two areas.

38 Kinesis — there is no directional stimulus, so the rate of movement or rate of turning is related to the environment. Woodlice in the damp area were moving less than those in the dry area, e.g. at 3 minutes, five woodlice were moving in the dry area and one woodlouse in the humid area.

39 With a null hypothesis that movement of the maggots is random, a third should go into each area (positive, negative and neutral). The χ^2 test is used to see if the results differ significantly from those expected from random movement. Plot a multiple-line graph of mean speed against time, using a different symbol for each sector. If there is no obvious pattern, count the numbers of times the speed increases and decreases with distance to see if there is a pattern.

40 15 out of 20 maggots showed negative phototaxis as they moved away from the light. They moved fastest when moving away from the light. In all cases the speed decreased with distance showing the speed may be related to light intensity.

41 Near stream, $D = 0.67$; away from stream, $D = 0.78$

42 Lower diversity near the stream; some species cannot tolerate lower salinity near the stream; two species are absent from this area. Salinity may be the abiotic factor limiting distribution of these species, especially *Gibbula umbilicalis*.

43 Null hypothesis: there is no significant difference between the observed and expected numbers of colour morphs on the stipes and tips of serrated wrack.
$\chi^2 = 18.92$; df = 3; $p < 0.001$, so the difference in numbers is significant
The null hypothesis is rejected.

44 The colour morphs of periwinkle are not settling at random on the serrated wrack; yellow are more concentrated at the tips, brown on the holdfasts and stipes. This may be to do with camouflage against predators, such as the common blenny, *Lipophyris pholis*, a fish that feeds on the periwinkle when the tide comes.